Annika Utz

The Electrochemical Oxidation of H_2 and CO at Patterned Ni Anodes of SOFCs

Schriften des Instituts für Werkstoffe der Elektrotechnik,
Karlsruher Institut für Technologie
Band 20

The Electrochemical Oxidation of H_2 and CO at Patterned Ni Anodes of SOFCs

by
Annika Utz

KIT Scientific Publishing

Dissertation, Karlsruher Institut für Technologie
Fakultät für Elektrotechnik und Informationstechnik, 2011

Impressum

Karlsruher Institut für Technologie (KIT)
KIT Scientific Publishing
Straße am Forum 2
D-76131 Karlsruhe
www.ksp.kit.edu

KIT – Universität des Landes Baden-Württemberg und nationales
Forschungszentrum in der Helmholtz-Gemeinschaft

KIT Scientific Publishing 2011
Print on Demand

ISSN 1868-1603
ISBN 978-3-86644-686-1

The Electrochemical Oxidation of H_2 and CO at Patterned Ni Anodes of SOFCs

Zur Erlangung des akademischen Grades eines

DOKTOR-INGENIEURS (DR.-ING.)

von der Fakultät für

Elektrotechnik und Informationstechnik

des Karlsruher Instituts für Technologie (KIT)

genehmigte

DISSERTATION

von

Dipl.-Ing. Annika Utz

geb. in: Mutlangen

Tag der mündlichen Prüfung: 04.02.2011

Hauptreferentin: Prof. Dr.-Ing. Ellen Ivers-Tiffée

Korreferentin: Prof. Dr. rer. nat. Dagmar Gerthsen

Danksagung

Die vorliegende Doktorarbeit entstand während meiner Forschungstätigkeit am Institut für Werkstoffe der Elektrotechnik (IWE) des Karlsruher Instituts für Technologie (KIT). An erster Stelle gilt mein Dank der Institutsleiterin Frau Prof. Ellen Ivers-Tiffée für die Betreuung und Förderung dieser Arbeit. Bei Frau Prof. Dagmar Gerthsen bedanke ich mich für das Interesse an meiner Arbeit und die Übernahme des Korreferats.

Für die erfolgreiche Zusammenarbeit im Rahmen des Projektes 'Mechanismus der Elektrooxidation von Reformatgasen an SOFC-Modellanoden' bedanke ich mich bei Frau Dr. Heike Störmer und den Herren Dr. Wolfgang Bessler, Dr. Marcel Vogler und Dr. Vitaliy Yurkiv, sowie bei der Deutschen Forschungsgemeinschaft für die Finanzierung meiner Stelle.

Die Untersuchung der Modellanoden auf Verunreinigungen erfolgte während meines Auslands-Forschungsaufenthaltes am Risø DTU National Laboratory for Sustainable Energy in Roskilde, Dänemark. Mein Dank gilt Herrn Dr. Mogens Mogensen und seinen Mitarbeitern für ihre Unterstützung und das angenehme Arbeitsklima. Für die finanzielle Unterstützung bedanke ich mich beim Karlsruhe House of Young Scientists (KHYS) am Karlsruher Institut für Technologie und dem Strategic Electrochemistry Research Center (SERC) am Risø DTU.

Meinen Kollegen am IWE danke ich für ihre Hilfsbereitschaft, die vielen interessanten Diskussionen und das kollegiale Arbeitsklima.

Ein besonderer Dank gilt meinen Eltern, meinen Geschwistern und Oliver für ihre großartige Unterstützung auf die ich mich immer verlassen kann.

Annika Utz
Stuttgart, im Mai 2011

Zusammenfassung

In den letzten Jahren besteht zunehmend Interesse an der Festoxidbrennstoffzelle (engl: Solid Oxide Fuel Cell, SOFC), die eine effiziente und emissionsarme Stromerzeugung bei gleichzeitiger Flexibilität bei der Wahl des Brennstoffs ermöglicht. Für eine breite kommerzielle Nutzung sind jedoch die Punkte Leistungsverbesserung, Langzeitstabilität und Kostenreduktion entscheidend. Es besteht insbesondere Verbesserungsbedarf mit Hinblick auf die Verlustprozesse an der Anode.

Grundvoraussetzung für die Verbesserung der Anodenleistung stellt dabei die Kenntnis der elementarkinetischen Reaktionsschritte der elektrochemischen Oxidation des Brenngases (H_2, CO oder Reformatgase) dar. Diese Reaktion findet an der Drei-Phasen-Grenze (engl: Triple-Phase-Boundary, TPB) von Elektrode, Elektrolyt und Gasraum statt. Dabei sind Untersuchungen an sogenannten Modellanoden mit zweidimensionaler Geometrie vorteilhaft, denn zusätzlich zur Vereinfachung der Geometrie ist eine Entkopplung der Oxidation von weiteren Verlustmechanismen in der porösen Struktur technisch relevanter Ni/8YSZ Cermetanoden möglich.

Gegenstand dieser Arbeit ist die detaillierte elektrochemische Charakterisierung von strukturierten Ni-Anoden (Modellanoden) sowohl in H_2-H_2O als auch CO-CO_2 Atmosphäre. Die mittels Impedanzspektroskopie ermittelten Parameterabhängigkeiten des linienspezifischen Widerstands (engl: Line Specific Resistance, LSR) stellen für jede der beiden Atmosphären einen umfassenden Datensatz dar. Die Stärke der Datensätze liegt dabei in den begleitenden, umfangreichen Untersuchungen zur Stabilität der Modellanoden sowie der Reproduzierbarkeit der Messergebnisse, ergänzt durch eine detaillierte Untersuchung auf Verunreinigungen der Drei-Phasen-Grenze. Die erstmals mögliche Vergleichbarkeit des Anodenverhaltens in den beiden Gasatmosphären erlaubt zusätzlich wertvolle Einblicke in den Reaktionsmechanismus und deutet auf ähnliche Reaktionsmechanismen hin. Der Vergleich zu technisch relevanten Cermetanoden mittels eines Kettenleitermodels zeigt eine große Ähnlichkeit der beiden Anodenkonzepte sowohl für die absoluten flächenspezifischen Widerstände (engl: Area Specific Resistance, ASR) als auch für die Parameterabhängigkeiten. Damit wird die Anwendbarkeit von Modellanoden zur Untersuchung der elektrochemischen Oxidation des Brennstoffs an SOFC-Anoden abschließend bestätigt.

Im Einzelnen wurden die folgenden, maßgeblichen Ergebnisse erzielt:

Entwicklung eines Ersatzschaltbildmodells

Eine hochauflösende Messdatenanalyse ermöglichte die Identifizierung von drei Verlustprozessen. Der größte Verlustprozess entsteht erwartungsgemäß durch den Ladungstransfer an der TPB der Modellanode ($P_{A\beta}$), die weiteren Prozesse sind auf Gasdiffusion ($P_{A\alpha}$) und die Cermet-Gegenelektrode ($P_{A\gamma}$) zurückzuführen. Dabei ist die Trennung der Verlustprozesse von Gasdiffusion und Ladungstransfer an der Modellanode besonders bei hohen Temperaturen entscheidend und ermöglicht erstmals die Bestimmung der Parameterabhängigkeiten des reinen Ladungstransferwiderstands LSR_{ct}.

Identifikation von Relaxations- und Degradationsmechanismen

Es wurde gezeigt, dass die Stabilität der Modellanodenstruktur sehr stark von der Schichtdicke, und den Auslagerungsbedingungen (insbesondere Temperatur) abhängt. Darüber hinaus wurden während der elektrochemischen Charakterisierung von nominell stabilen Modellanoden (Schichtdicke: 800 nm) mehrere Relaxations- und Degradationsmechanismen mit unerwartet großen Zeitkonstanten beobachtet:

(i) Anfängliche Abnahme von LSR_{ct} um 60 % mit einer Stabilisierung bei 54 Ω·m nach ca. 20 h bei 800 °C, verursacht durch eine Veränderung der Ni-Struktur.

(ii) Große, gasabhängige Degradationsraten, mit stärkerer Degradation in H_2-H_2O (0.4 % / h) als in CO-CO_2 (0.18 % / h), vermutlich aufgrund der Bildung von gasförmigem $Ni(OH)_2$.

(iii) Langsame Relaxationsprozesse nach Gasvariation von pH_2 und pH_2O mit Einschwingzeiten von 2 - 3 h bei pH_2 und 5 - 8 h bei pH_2O (800 °C), vermutlich aufgrund von langsamen Bulkprozessen. In CO-CO_2 Atmosphäre erfolgt die Stabilisierung von LSR hingegen innerhalb weniger Minuten.

(iv) Schnelle Zellaktivierung während der Aufzeichnung von Strom-Spannungs-Kennlinien, gefolgt von einem langsamen Relaxationsprozess während ca. 20 h.

Die reproduzierbare Charakterisierung von Modellanoden erfordert damit als Voraussetzung die standardisierte Vorbehandlung aller Zellen, sowie entsprechend lange Haltezeiten nach Variation der Gaszusammensetzung oder der Aufzeichnung von Strom-Spannungs-Kennlinien.

Ermittlung der Parameterabhängigkeiten von LSR in H_2-H_2O

Alle relevanten Parameterabhängigkeiten von LSR_{ct} in H_2-H_2O Atmosphäre für Variationen der Gaszusammensetzung (pH_2, pH_2O, β_{rel}), der Temperatur (T) und der Überspannung (η) wurden bestimmt:

(i) Die starke, negative Abhängigkeit von pH_2O (b = 0.68) bestätigt die Beobachtungen aus der Literatur.

(ii) Die Abhängigkeit von pH_2 ist gering.

(iii) Die Aktivierungsenergie ändert sich mit der Gaszusammensetzung mit Werten von E_{act} = 1.01 - 1.37 eV (T = 450 bis 800 °C).

(iv) Der Ladungstransferkoeffizient α = 0.61 bis 0.69 (T = 650 bis 800 ° C) wurde mit Impedanzspektren unter Polarisation ermittelt.

Der durch die Charakterisierung von Proben mit unterschiedlichem Design ermittelte lineare Zusammenhang von LSR_{ct} und l_{TPB} bestätigt, dass die Reaktion an der Drei-Phasen-Grenze stattfindet.

Darüber hinaus wurde der Einfluss der Doppelschichtkapazität an der Zwei-Phasen-Grenze (2PB) von Elektrode und Elektrolyt bei der Auswertung des kapazitiven Beitrags von Modellanoden berücksichtigt. Durch die Charakterisierung von Zellen mit stark unterschiedlichem 2PB/TPB Verhältnis war erstmals eine Aufteilung der Kapazität in einen linienspezifischen Anteil des Ladungstransfers und einen flächenspezifischen Anteil der Doppelschichtkapazität möglich. Die experimentell ermittelten Werte für die Doppelschichtkapazität zeigen eine gute Übereinstimmung zur theoretischen Abschätzung entsprechend der Theorie von Stern.

Ermittlung der Parameterabhängigkeiten von LSR in CO-CO_2

Entsprechend den Arbeiten in H_2-H_2O, wurden die Parameterabhängigkeiten von LSR in CO-CO_2 für Variationen von Gaszusammensetzung (pCO, pCO_2, β_{rel}) und Temperatur (T) bestimmt:

(i) Die starke negative Anhängigkeit von pCO_2 (d = 0.61 bis 0.79) zeigt große Ähnlichkeit zum Verhalten bei einer Variation von pH_2O.

(ii) Das Verhalten bei pCO Variation zeigt einen Knick bei pCO ≈ $2 \cdot 10^4$ Pa von negativer Abhängigkeit bei kleinerem pCO zu schwächerer negativer Abhängigkeit bzw. zu leicht positiver Abhängigkeit, je nach pCO_2.

(iii) Mit zunehmendem pCO ändert sich die Aktivierungsenergie E_{act} von 0.85 zu 1.42 eV.

Zudem wurde der lineare Zusammenhang von LSR_{ct} und l_{TPB} bestätigt.

Vergleich zu Cermetanoden mittels eines Kettenleitermodells

Der Vergleich der an Modellanoden erzielten Ergebnisse zu technisch relevanten Cermetanoden wurde durch die Verwendung eines Kettenleitermodells realisiert. Ein solches Modell kombiniert die Verlustprozesse von Ladungstransfer an der TPB und von ionischem Transport im Elektrolyten. Dabei müssen Annahmen für die Modellparameter volumenbezogene TPB-Länge ($l_{TPB,v}$) und effektive ionische Leitfähigkeit im Elektrolyten ($\sigma_{ion,eff}$) getroffen werden. Die Berechnung zeigt eine sehr gute Übereinstimmung der beiden Anodenkonzepte mit Hinblick auf absolute ASR Werte und die Aktivierungsenergie. Bei den Abhängigkeiten von der Gaszusammensetzung zeigt sich eine gute qualitative Übereinstimmung, allerdings ergeben sich jeweils stärkere theoretische Abhängigkeiten als die experimentell an Cermetanoden bestimmten Werte. Dies kann einerseits auf die Parameterschätzung zurückzuführen sein, andererseits aber auch ein Indikator für einen zusätzlichen Verlustmechanismus an der TPB von Modellanoden sein, beispielsweise durch Verunreinigungen.

Rückschlüsse zum Reaktionsmechanismus

Die Ähnlichkeit der Ergebnisse in Bezug auf absolute LSR_{ct} Werte und Parameterabhängigkeiten, die bei der elektrochemischen Charakterisierung in H_2-H_2O und CO-CO_2-Atmosphäre ermittelt wurden, deuten auf ähnliche Reaktionsmechanismen in beiden Gasatmosphären hin. Im Fall von H_2-H_2O wird dabei meist ein H-Spillover Mechanismus als ratenbestimmender Schritt angenommen. Interstitieller Ladungstransfer über die Zwei-Phasen-Grenze wird im CO-CO_2 Betrieb aufgrund der Molekülgröße als unwirksam eingeschätzt. Im Fall von H_2-H_2O könnte dieser Mechanismus jedoch teilweise zur Stromdichte beitragen und damit die beobachteten, langsamen Relaxationsprozesse im H_2-H_2O Betrieb erklären.

Obwohl aufgrund der Parameterschätzung für das Kettenleitermodell ein zusätzlicher geringer Verlustmechanismus an der TPB von Modellanoden - beispielsweise durch Verunreinigungen - nicht ausgeschlossen werden kann, deutet die gute Übereinstimmung der Leistungen von Modell- und Cermet-Anoden auf den gleichen zugrundeliegenden Reaktionsmechanismen hin. Damit bestätigt diese Arbeit die Anwendbarkeit von Modellanoden zur Untersuchung der elektrochemischen Oxidation des Brenngases an SOFC-Anoden.

Table of Contents

1 Introduction

In the past few years, interest in Solid Oxide Fuel Cells (SOFC) has increased significantly. Their advantages include an efficient and low emission power generation as well as fuel flexibility [1]. However, major challenges on the way to a commercial launch include performance improvement, long-term stability and cost reduction. With respect to overall performance improvements, a recent comparison of electrode performance clearly indicates the necessity to improve the anode side of SOFCs [2].

The knowledge of the elementary kinetic steps of the electrochemical oxidation of fuel gas at the anode is a basic requirement for the improvement of anode performance [3]. This reaction takes place at the so-called Triple-Phase-Boundary (TPB), where electrode, electrolyte and gas phase join. So far, several reaction mechanisms have been proposed for the charge transfer reaction [4], but still the exact mechanism is unclear. For a deeper understanding of the reaction mechanism, various simplified geometries of model anodes have been established. This includes point and patterned Ni anodes with two-dimensional TPBs.

In order to draw conclusions on the rate determining reaction mechanism, an experimental data basis of the anodic activation polarization is required in combination with elementary kinetic modeling studies. Such a data basis includes the Line Specific Resistance (LSR) attributed to charge transfer and its dependencies on gas composition, temperature and polarization voltage. Although several studies have been performed so far with patterned Ni anodes in H_2-H_2O atmosphere, these turn out to be either inconsistent or incomplete [5-9]. For instance, considerable discrepancies between the different studies exist for the LSR and the respective parameter dependencies and no information is given on reproducibility and impurities. Furthermore, the evaluation of impedance spectra is typically performed without a separation of the different loss mechanisms.

Even less information exists on the electrochemical characterization of SOFC anodes in CO-CO_2 atmosphere and in reformate gases. However, these fuel gases are of special interest for SOFC operation with a variety of fuels.

In this work, a detailed experimental characterization of patterned Ni anodes will be performed in H_2-H_2O and CO-CO_2 atmosphere. With respect to the observed discrepancies in previous studies in literature, emphasis will be on sample stability, measurement reproducibility and impurity analysis. High resolution data analysis will be used to separate charge transfer from other loss mechanisms. Furthermore, this work allows a comparison of patterned Ni electrode performance in the two gas atmospheres as well as a comparison to point anode and cermet anode performance and thus contributes to a deeper understanding of the reaction mechanisms of the electrochemical oxidation of the fuel gas at SOFC anodes.

1.1 Introduction to Solid Oxide Fuel Cells

In this section, the key features of SOFCs which are necessary for the understanding of this work are introduced. A special focus will be directed to the anode and its microstructure.

1.1.1 Working Principles

The main component of a Solid Oxide Fuel Cell (SOFC) is the gas tight but oxygen-ion conducting electrolyte. It separates the two gas chambers that are continuously supplied with the corresponding gases: fuel gas (e.g. H_2, CO, reformate gases) at the anode side, and oxygen or air at the cathode side, see figure 1.1a. The strong difference in chemical potential between the high oxygen partial pressure at the cathode and a very low oxygen partial pressure at the anode is the driving force for oxygen ion diffusion through the electrolyte.

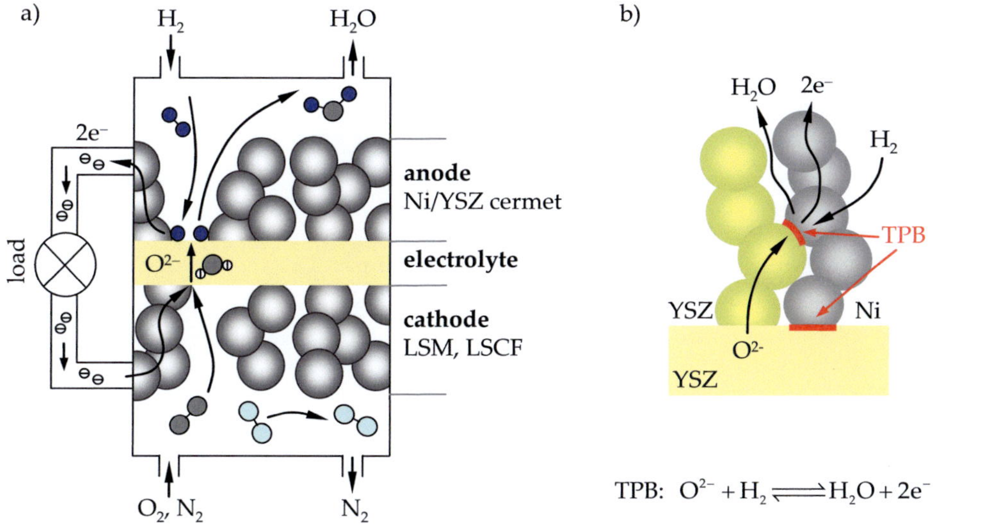

figure 1.1: a) Scheme of the functioning of SOFCs: oxygen and fuel (here: H_2) react via a dense, oxide ion-conducting electrolyte. The spatial separation of reduction and oxidation reaction enables the utilization of the electrons involved in the redox process, and b) schematic concept of a Ni/8YSZ cermet anode: the electrochemical oxidation of H_2 takes place at the Triple-Phase-Boundary (TPB), where electrode, electrolyte and gas phase join.

At the cathode, the O_2 is reduced to O^{2-} by taking up two electrons and is incorporated into the electrolyte in a subsequent step, see equation 1.1:

$$\tfrac{1}{2}O_2(\text{gas}) + 2e^- \rightleftharpoons O^{2-}(\text{electrolyte}) \qquad 1.1$$

At the anode, H_2 is oxidized with O^{2-}, forming gaseous water and releasing two electrons:

$$O^{2-}(\text{electrolyte}) + H_2(\text{gas}) \rightleftharpoons H_2O(\text{gas}) + 2e^- \qquad 1.2$$

The overall reaction at SOFC fuel cells driven with $H_2 - H_2O$ atmosphere is then:

$$\tfrac{1}{2}O_2(\text{gas}) + H_2(\text{gas}) \rightleftharpoons H_2O(\text{gas}) \qquad 1.3$$

At open circuit, when electrons are not able to move back to the cathode, an electric field force is established and impedes further oxygen ion diffusion through the electrolyte. In the case of an ideal, reversible SOFC, this Nernst voltage can be expressed using the Gibbs free energy for the oxidation of the fuel gas [10]:

$$U_{Nernst} = -\frac{\Delta G_{0,H_2-H_2O}(T)}{n_e F} - \frac{RT}{n_e F}\ln\left(\frac{pH_2O_{an}}{\sqrt{pO_{2,cat}}\,pH_{2,an}}\right) \qquad 1.4$$

Here $\Delta G_{0,H2-H2O}$ is the Gibbs free energy of the H_2 oxidation reaction, T is the absolute temperature, n_e the number of electrons involved in the reaction (n_e = 2 in the case of SOFC), R is the universal gas constant, and F the Faraday constant.

In the case of an ideal, loss-free functioning of the fuel cell, the change in Gibbs free energy of the reaction is completely converted into electrical work. The maximum obtainable Nernst voltage is a function of the atmosphere (e.g. H_2-H_2O or CO-CO_2), the gas composition and temperature.

In real SOFCs at load, the voltage will be lowered compared to the thermodynamically predicted value U_{Nernst} as the cell performance is limited by three loss mechanisms namely (i) ohmic losses, (ii) polarization losses due to gas diffusion and (iii) losses by activation polarization.

1.1.2 Microstructure of the Anode

The electrochemically active area of the SOFC anode is restricted to the Triple-Phase-Boundary (TPB), where electrode, electrolyte and gas phase join, see figure 1.1b. Only at these points, the reactants for the oxidation reaction (i.e. oxygen ions and fuel gas) are supplied and the reaction products (i.e. electrons and water vapor) are removed. Thus, the anode performance increases with increasing amount of TPB. Therefore, the quantity l_{TPB} is introduced to indicate the TPB length per electrode area. By the fabrication of a porous mixture of electrode and electrolyte particles, large values of l_{TPB} can be realized in the electrode volume.

State of the art technical anode structures are composed of a combination of ceramic electrolyte particles and metallic electrode particles, denoted as cermet (short for Ceramic Metal) where the notation Ni/8YSZ indicates the materials: for the electrode, the catalytically active and comparatively inexpensive Nickel is used, and for the electrolyte 8mol% Y_2O_3-stabilized ZrO_2 (8YSZ) is used. Considering the material composition, the percolation of the ionic and the electronic path as well as the gas phase must be maintained. Furthermore, the addition of electrolyte material to the anode also allows for a stabilization of the porous electrode structure by preventing the Ni grains from sintering together and by matching the thermal expansion coefficients of both materials [1].

In addition to the activation polarization losses, another loss mechanism arises due to gas diffusion in the porous cermet structure. In order to keep both polarization losses as low as possible, an anode structure composed of two layers with different properties is advantageous [11, 12]:

(i) In the layer close to the electrode-electrolyte interface, a high value of l_{TPB} is realized by small particles and low porosity of about 20 %. Since the electrochemical oxidation takes place in this layer, it is denoted Anode Functional Layer (AFL).

(ii) In the outermost layer of the electrode, sufficient gas supply is enabled by a high porosity of the substrate of about 40 %. This layer is referred to as anode substrate.

In the case where the single cell is built on an anode substrate, the concept is denoted Anode Supported Cell (ASC). Compared to Electrolyte Supported Cells (ESC), where the fuel cell is supported by the electrolyte, ASCs allow to reduce the ohmic losses in the electrolyte since very thin layers can be realized.

The activation polarization at the anode is attributed to the reaction mechanisms at the TPB and represents a Line Specific Resistance (LSR) with unit Ω·m, which is a function of gas composition, temperature and polarization resistance. It is determined experimentally based on the polarization resistance attributed to charge transfer (R_{ct}) multiplied with the TPB length per electrode area (l_{TPB}) and the electrode area (A):

$$LSR_{ct} = R_{ct} \cdot l_{TPB} \cdot A \qquad 1.5$$

However, in technical cermet anodes the TPB length is not well known and the polarization resistance includes an additional contribution of ionic transport in the YSZ matrix. Therefore, the electrode performance of technical cermet anodes is described by an Area Specific Resistance (ASR) that combines the contributions of charge transfer and ionic transport.

1.1.3 Kinetics

In elementary kinetic studies, electrochemical reactions such as the anode reaction in equation 1.2 are further divided into individual reaction steps [13]. These individual reaction steps include the electron transfer reaction (typically referred to as charge transfer), heterogeneous chemistry as well as surface and bulk diffusion.

For the charge transfer reaction, different steps have been suggested and discussed in literature, see figure 1.2. An overview is given by Adler and Bessler in [4]. These mechanisms include charge transfer at the TPB during hopping of:

(i) adsorbed hydrogen from Ni to YSZ, denoted H spillover (figure 1.2a)

(ii) oxygen ions from YSZ to Ni, denoted O spillover (figure 1.2b)

(iii) hydroxyl either from Ni to YSZ or inverse, denoted OH spillover (figure 1.2c).

Additionally, the possibility of charge transfer at the Two-Phase-Boundary (2PB) of electrode and electrolyte by interstitial transport of atomic hydrogen is proposed, see figure 1.2d.

a)

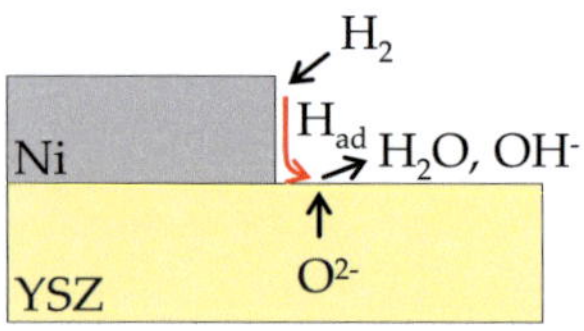

$$H_{Ni} + O^{2-}_{YSZ} \rightleftharpoons OH^-_{YSZ} + e^-_{Ni} + \square_{Ni} \quad 1.6$$

$$H_{Ni} + OH^-_{YSZ} \rightleftharpoons H_2O_{YSZ} + e^-_{Ni} + \square_{Ni} \quad 1.7$$

b)

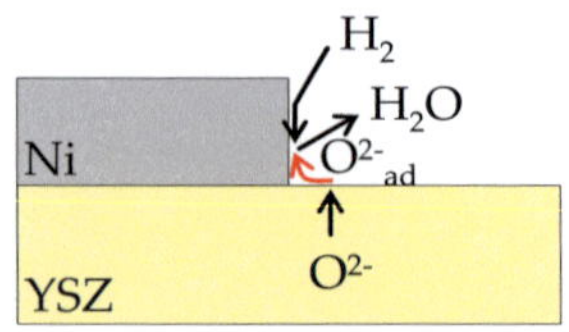

$$O^{2-}_{YSZ} + \square_{Ni} \rightleftharpoons O^-_{Ni} + e^-_{Ni} + \square_{YSZ} \quad 1.8$$

$$O^{2-}_{YSZ} + \square_{Ni} \rightleftharpoons O_{Ni} + 2 \cdot e^-_{Ni} + \square_{YSZ} \quad 1.9$$

$$O^{2-}_{YSZ} \rightleftharpoons O^-_{YSZ} + e^-_{Ni} \quad 1.10$$

$$O^-_{YSZ} + \square_{Ni} \rightleftharpoons O_{Ni} + e^-_{Ni} + \square_{YSZ} \quad 1.11$$

c)

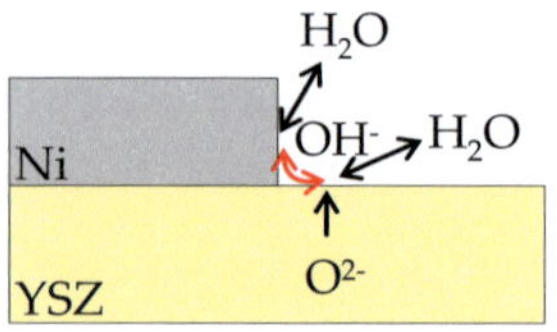

$$OH^-_{YSZ} + \square_{Ni} \rightleftharpoons OH_{Ni} + e^-_{Ni} + \square_{YSZ} \quad 1.12$$

d)

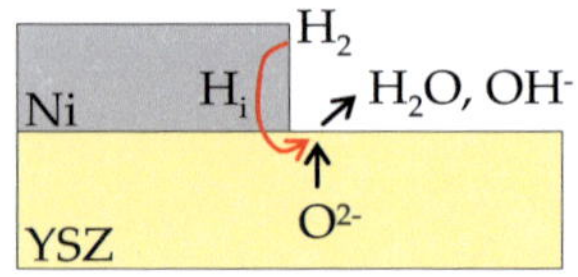

$$H^{x}_{i,Ni} + V_{YSZ} \rightleftharpoons H^{\bullet}_{i,YSZ} + e^-_{Ni} + V_{Ni} \quad 1.13$$

figure 1.2: Scheme and equations of different elementary kinetic reaction mechanisms for the electrochemical oxidation of hydrogen: a) charge transfer during hopping of surface adsorbed H_{Ni} from Ni to YSZ surface with reaction with O^{2-} to H_2O_{YSZ} or OH^- also denoted hydrogen spillover, b) charge transfer during hopping of O^{2-} or O^- from YSZ to the Ni surface also denoted oxygen spillover, c) charge transfer during hopping of OH^- from YSZ to Ni or Ni to YSZ, also denoted hydroxyl spillover, d) charge transfer between interstitial hydrogen H_i at the electrode-electrolyte interface, denoted interstitial charge transfer. Here, □ indicates a free surface site, V indicates a bulk vacancy and the subscripts Ni and YSZ denote the respective surface or bulk.

Herein, a distinction is made between single and two step mechanisms. In the former, the two electrons of an O^{2-} ion are transferred at once (e.g. equation 1.9), in the latter the two electrons are transferred in two consecutive steps (e.g. following equations 1.10 and 1.11).

The reaction mechanisms of figure 1.2a-c apply for clean surfaces of electrode and electrolyte and especially a clean TPB. In the case of impurity segregations at the TPB, the reaction mechanisms may be subject to a blocking and a change in the reaction mechanism e.g. in favor of interstitial charge transfer, see figure 1.2d.

In the case of CO-CO_2 atmosphere, only sparse literature exists. Mostly, charge transfer mechanisms with oxygen spillover are considered. Especially the interstitial pathway seems improbable, since solubility and diffusivity of CO in Ni are assumed to be very low [14].

In most studies, the charge transfer step is denoted according to the accompanying transfer mechanism of either Ni-surface adsorbed H_{Ni} from Ni to YSZ or YSZ-surface adsorbed $O^{2-}{}_{YSZ}$ from YSZ to Ni. These are typically referred to as spillover mechanisms [4, 15, 16]. This expression is borrowed from heterogeneous catalysis, where it is defined as the transport of an activated species from one surface to another surface on which it cannot adsorb under similar conditions [17]. While the applicability of this definition for the electrochemical oxidation at the SOFC is argued [15, 18], the common denotation as spillover mechanisms is maintained throughout this work.

1.1.4 Model Anodes

Due to the complex structure of cermet anodes, model anodes are increasingly used for experimental and theoretical investigations of the anode reaction mechanisms. These anodes exhibit a two-dimensional and thus easily accessible l_{TPB}. Among them are micropatterned electrodes, point electrodes or grid electrodes, see

figure 1.3. Furthermore, the simplified geometry allows to separate the elementary kinetic charge transfer reaction at the TPB from the role of the anode morphology.

Among the different model anode concepts, patterned Ni anodes enable the reproducible preparation of comparatively high values of l_{TPB} up to several meters. With higher values of l_{TPB}, parasitic influences are reduced in favor of the measurement accuracy.

Even though patterned Ni anodes represent a simplified anode structure, results from electrochemical characterization indicate difficulties with respect to the reproducibility of electrochemical data [6-9]: the comparison of results from different studies in literature shows significant deviations in absolute values of LSR, as well as controversial parameter dependencies.

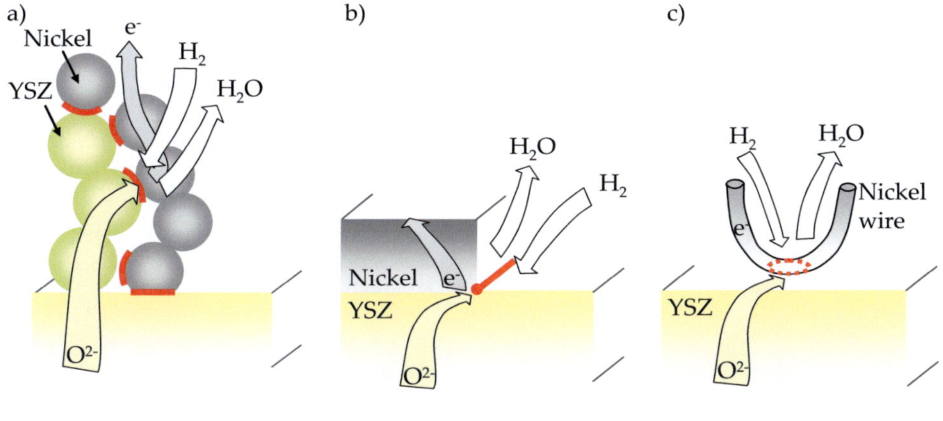

figure 1.3: Scheme of the reaction mechanism at different anode concepts: a) cermet structure with long transport paths for oxygen ions, electrons and fuel gas, b) micropatterned Ni anode with well defined TPB, and c) point anode prepared by bending and pressing a Ni wire on the electrolyte substrate.

1.2 Goals of this Work

The aim of this work is to establish a complete data basis of the electrochemical characterization of patterned Ni anodes in H_2-H_2O and CO-CO_2 atmosphere, including the parameter dependencies of the Line Specific Resistance (LSR) on gas composition, temperature and polarization voltage.

With respect to the persisting discrepancies in literature, special focus will be directed to (i) the stability of the Ni structure and (ii) the measurement reproducibility, as well as (iii) the combined analysis of impedance spectra with Distribution of Relaxation Times (DRT) calculation and Complex Nonlinear Least Squares (CNLS) fitting in order to separate the contribution of charge transfer from other contributions. Furthermore, the possibility of an influence of impurities on the reaction mechanisms will be considered by a detailed surface analysis and the identification of impurity features.

By the comparison of the obtained LSR values and parameter dependencies on gas composition and temperature for the two analyzed atmospheres, a deeper understanding of the reaction mechanism is expected.

Furthermore, a comparison of different anode concepts (patterned -, cermet -, and point anodes) will be performed to gain valuable insight on the elementary kinetics: the comparison of different model anodes with known impurity content gives information on a possible influence of impurities. Finally, the comparison to experimental data of technical Ni/8YSZ cermet anodes by means of a transmission line model gives evidence whether the geometry simplifications applied for model anode studies are valid. Hence, the applicability of model anodes for the identification of the elementary kinetic reaction mechanism at the SOFC anode is analyzed.

1.3 Outline

In the following chapter 2, a detailed review of previous patterned Ni anode studies dealing with (i) the electrochemical characterization and (ii) the modeling of the elementary kinetics in H_2-H_2O and CO-CO_2 atmosphere is given, as well as (iii) the state of the art of impurity analysis studies. The main characteristics of patterned Ni anodes – fabrication, stability, electrochemical characterization and data analysis, and impurity features – are addressed in chapter 1.

The subsequent chapters present the results of electrochemical characterization of the patterned Ni anodes:

(i) In chapter 4, the degradation and relaxation mechanisms observed during the characterization of nominally stable patterned Ni anodes are described and discussed.

(ii) In chapter 5, the results of the electrochemical characterization in H_2-H_2O atmosphere are given together with the comparison to data from patterned Ni anode studies in literature and to data from point anodes with known impurity features.

(iii) In chapter 6, the results of the electrochemical characterization in CO-CO_2 atmosphere are given and a comparison of the electrode performance in H_2-H_2O and CO-CO_2 is discussed with respect to possible reaction mechanisms.

(iv) In chapter 7, the comparison to electrochemical performance of technical Ni/8YSZ cermet anodes is performed using a transmission line model and the results are discussed.

A concluding discussion of the achievements of this work is given in chapter 8.

2 Evaluation of Previous Studies

This chapter summarizes the state of the art in both electrochemical characterization and elementary kinetic modeling of SOFC anodes with special emphasis on patterned Ni anodes.

The main step in order to obtain evidence on the rate determining reaction mechanism consists of combined studies on electrochemical characterization of simplified geometries and on elementary kinetic modeling. Herein, the dependencies of LSR on pH_2, pH_2O, pCO, pCO_2, T, and η are important parameters. Additionally, other effects such as the influence of impurities give valuable insight.

An overview of previous studies on the oxidation reaction of H_2 and CO with patterned Ni anodes is given (i) with respect to the electrochemical characterization in section 2.1, and (ii) with respect to the elementary kinetic modeling in section 2.2. The findings of the studies on the impurity analysis of Ni point anodes are summarized in section 2.3.

2.1 Electrochemical Characterization

2.1.1 H_2-H_2O Atmospheres

A lot of studies exist for the electrochemical characterization of SOFC anodes in H_2-H_2O atmosphere. These include model anodes – mainly point anodes [19-26] and patterned anodes [5-9, 27-30] – as well as technical Ni/8YSZ cermet anodes [31-35]. With respect to the large number of studies and the scope of this work, only the electrochemical performance of patterned Ni anodes is discussed in detailed. Further aspects of the characterization that have not been reported for patterned anodes are summarized briefly.

Patterned Ni Anodes

Standard literature data for patterned Ni anodes comprises the studies by three groups with comparable experimental conditions [6-9]: patterned Ni anodes with a layer thickness around 1 µm, l_{TPB} between 0 to 12 $m \cdot cm^{-2}$, smallest stripe width between 5 to 10 µm, three-point measurement setup with Pt reference electrodes, and single gas atmosphere. The parameter studies include variations of pH_2 and pH_2O in different ranges at temperatures between 400 to 850 °C. However, a comparative analysis of the investigated gas compositions reveals almost no coincident points, as shown in figure 2.1. Furthermore, the range of relative water vapor content (β_{rel})

is different. The relative water vapor content is calculated following $\beta_{rel} = pH_2O / (pH_2O + pH_2)$. As a consequence, the existing experimental data can only be compared at around $pH_2 \approx 10^4$ Pa and $pH_2O \approx 10^3$ Pa.

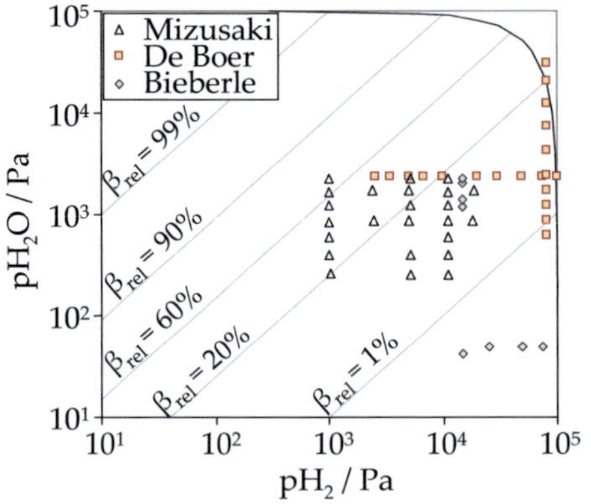

figure 2.1: Comparison of experimental conditions of pH_2 and pH_2O variation in the studies with patterned Ni anodes by Mizusaki [8, 9], de Boer [7] and Bieberle [6]. β_{rel} indicates the relative water vapor content, the bold line indicates atmospheric pressure.

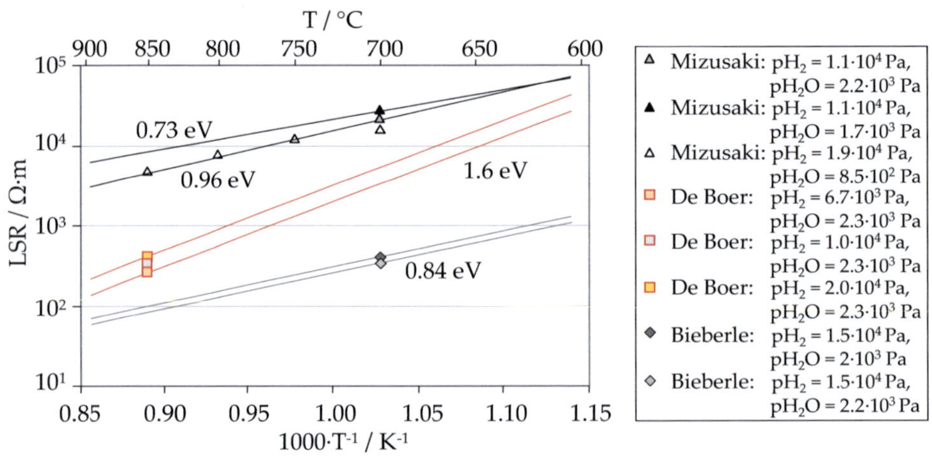

figure 2.2: Comparison of LSR values reported in literature [6-9] at a gas composition of $pH_2 \approx 10^4$ Pa and $pH_2O \approx 10^3$ Pa together with the activation energy behavior.

table 2.1: Comparison of LSR values reported in literature [6-9]. The values are recorded at similar gas compositions and projected to 700 and 850 °C using the respective activation energy E_{act}.

Study	Ref.	LSR [Ω·m]	T [°C]	pH_2O [Pa]	pH_2 [Pa]	β_{rel} [%]	$l_{TPB}\cdot A$ [m]	E_{act} [eV]	LSR_{calc}(700 °C) [Ω·m]	LSR_{calc}(850 °C) [Ω·m]
Mizusaki	[8, 9]	20419	700	$2.2\cdot10^3$	$1.1\cdot10^4$	17.0	3.64	0.75	20419	6182
De Boer	[7]	344	850	$2.3\cdot10^3$	$9.9\cdot10^3$	18.8	8.31	1.60	4400	344
Bieberle	[6]	349	700	$2.2\cdot10^3$	$1.5\cdot10^4$	12.7	3.65	0.88	349	86

The comparative study of data gathered from samples with different pattern designs is enabled by the line specific resistance (LSR). The LSR is calculated as the product of polarization resistance (R_{pol}), TPB length per electrode area (l_{TPB}) and electrode area (A). So far, only values for the polarization resistance (R_{pol}) have been reported in literature [6-9]. In this work, the absolute LSR values have been calculated by extracting the reported values for R_{pol}, l_{TPB} and A from the respective studies. The resulting LSR values are shown in figure 2.2. An additional summary of the values projected to 700 and 850 °C using the reported activation energies (E_{act}) is given in table 2.1. The deviation is significant, with LSR values differing by almost two orders of magnitude for both temperatures. Further disagreements concern (i) the number of arcs observed in the impedance plot, (ii) the activation energies (E_{act}) reported, as well as (iii) the LSR dependencies on pH_2O and pH_2 (including a controversial behavior upon pH_2 variation), as summarized in table 2.2. Data by Mizusaki further indicates a difference in gas composition dependency upon temperature changes [9].

The studies reported neither (i) results from an impurity analysis, nor (ii) the exact measurement program, nor (iii) the measurement reproducibility. De Boer suspected a correlation between previous measurements and momentary electrode resistance, but no further analysis has been performed and no information is given with respect to the exact conditions of data acquisition [7].

table 2.2: Comparison of experimental results reported in literature [6-9]. Values a and b describe the dependencies of LSR on pH_2 and pH_2O respectively, according to the expression $LSR = c \cdot pH_2^{-a} \cdot pH_2O^{-b}$.

Study	Ref.	# EIS arcs	a	b	E_{act} [eV]
Mizusaki	[8, 9]	1	-0.09 to 0.15	0.32 to 0.88	0.75
De Boer	[7]	3	-0.26	0.37	1.60
Bieberle	[6]	1	0.11	0.67	0.88

Further studies indicated problems encountered in the characterization of patterned anodes, even though this has not been at the focus of the respective analysis: Sukeshini [29] and Ehn [27] reported that patterned Ni anodes have ended up as separate Ni islands with an undefined TPB length after testing. This result clearly demonstrates that structural changes during the characterization need to be taken into account. Unfortunately, in both studies the course of the polarization resistance with time was not measured and a variation of pH_2, pH_2O and T has not been performed. In a recent study, Ehn reported on the degradation behavior of two different types of patterned Ni anodes (dense and porous) [36]. However, for the determination of the parameter dependencies, pH_2 and pH_2O have not been varied independently.

For stability reasons, Yamamura [30] used two different samples for the characterization in the range of 500 – 750 °C and 700 – 800 °C. A comparison of the polarization resistances at 700 °C and 750 °C shows deviations of up to one order of magnitude. Unexpectedly, the higher resistance is reported for the sample characterized at the higher temperature. Thus, the difference in performance can not be attributed to a pronounced grain growth and higher l_{TPB} at elevated temperatures, but hints at a degradation process.

So far, only one study has addressed the influence of the electrolyte on the electrochemical performance of patterned Ni anodes. Using YSZ electrolytes with different crystal orientations, Rao et al. [37] have determined capacitances between 200 - 350 $\mu F \cdot cm^{-2}$ at 550 °C. This value combines the capacitive contributions of charge transfer and double layer. However, no separation of the distinct contributions has been performed. In general, literature dealing with the double layer capacitance at solid-solid interfaces is sparse [38-41].

Rate Determining Reaction Mechanism

When it comes to the interpretation of the rate determining reaction step, the expected linear correlation was found between polarization resistance and l_{TPB} [6, 7, 9], indicating that the reaction takes place at the TPB. As rate determining step, Mizusaki [9] suggested either dissociative adsorption of H_2 or surface diffusion of H_{ad} as rate determining step, whereas Bieberle et al. suggested either adsorption/desorption of H_2 or removal of O^{2-} from the electrolyte [6].

For a deeper understanding of the reaction mechanism and especially the role of electrode and electrolyte, several additional experiments have been reported. However, they were not performed with patterned anodes but with Ni/8YSZ cermet anodes: Primdahl reports a temperature dependent isotope effect upon a change from H_2-H_2O to D_2-D_2O [42]. This is interpreted as a shift in the rate determining step where at low temperature either a transport limitation or a limitation by a H_2 species reaction are suggested. The observed significant effect of the metal on electrode performance with best performance for Ni indicates that the rate determining reaction involves the Ni electrode [18]. Primdahl comes to the same interpretation with data of a sulfur poisoning experiment: the increase in resistance with H_2S addition to the fuel gas is explained by a blocking of Ni surface sites. However, Sasaki showed that Sc-doped ZrO_2 (ScYSZ) is less affected by sulfur poisoning [43], which rather indicates a reaction mechanism involving either reaction on or diffusion through the electrolyte [18, 43].

2.1.2 CO-CO_2 Atmospheres

Studies with CO-CO_2 atmosphere are of interest (i) for the comparison of the performance to data obtained in H_2-H_2O atmosphere, which allows conclusions on the kind of the reaction mechanism, and (ii) because the common fuel for SOFC is reformate gas that – upon internal reforming – splits up into CO components. Hence, the electrochemical characterization of the pure CO-CO_2 system can be considered as intermediate step towards the characterization with reformate gases.

A general discussion is whether different mechanisms apply for the electrochemical oxidation in H_2-H_2O and CO-CO_2 atmospheres: whereas several studies state that the anodic reaction is not affected by the kind of fuel but only by the partial pressure of oxygen (pO_2) and hence the reaction mechanisms are similar in both atmospheres [44, 45], other studies report a 2 – 10 times higher polarization resistance for CO-CO_2 atmospheres compared to H_2-H_2O [14, 29, 46-50]. In the following, an overview of existing literature dealing with the electrochemical characterization of Ni anodes in CO-CO_2 atmosphere is given.

Patterned Ni Anodes and Model Anodes

So far, literature on the electrochemical characterization in CO-CO_2 with patterned Ni anodes is sparse. Studies by Sukeshini et al. [29] as well as Ehn et al. [27] were performed with rather thin electrode layers of 200 nm and 0.5 – 1 μm thickness, respectively. In accordance with the findings of the stability study for Ni thin films performed in this work (see section 3.1.3 and [51]), theses structures proved to be unstable: bubbles formed under the Ni film or the continuous film even split up into separate islands during electrochemical characterization. Furthermore, no data is given on the polarization resistance as a function of the parameters pCO, pCO_2 and T. Hence, no quantitative parameter dependencies can be extracted from these studies. However, the studies by Sukeshini and Ehn report larger polarization resistances for CO-CO_2 than for H_2-H_2O atmosphere. A recent study by Ehn showed that the increase in polarization resistance for a shift from H_2-H_2O to CO-CO_2 is most pronounced for porous patterned Ni anodes, whereas dense electrodes show only a slight performance change [36]. This indicates an increase in polarization resistance due to an additional transport limitation rather than due to a change in reaction mechanism. Additionally, data by Sukeshini showed a dominance of H_2 oxidation during the oxidation of CO-H_2 mixtures [29].

Boulenouar et al. reported a study with Ni grid electrodes in CO-CO_2 atmosphere [47]: an exponent of d = 0.5 is reported for pCO_2 variation and of c = 0.5 for pCO variation at high pCO values. For the entire variation range of pCO a smaller exponent of c = 0.25 is obtained. Here, the exponents c and d are determined in accordance to a and b in H_2-H_2O atmosphere, using the expression $LSR = c \cdot pCO^{-c} \cdot pCO_2^{-d}$. No values are given for the activation energy.

In a combined experimental and modeling study, Lauvstad et al. characterized Ni and Pt point anodes in CO-CO_2 atmosphere [49]. Neither the reaction orders nor the activation energy are reported, as too few operation points were used for the determination of the quantitative dependencies. However, an increase in polarization resistance with increasing pCO (thus negative value of c) and a decrease with increasing pCO_2 (thus positive value of d) were observed.

Ni/8YSZ Cermet Anodes

A comprehensive study on Ni/8YSZ cermet anodes has recently been performed by Leonide et al. and reports the parameter dependencies for pCO, pCO_2 and temperature variations [52, 53]. The dependencies for pCO (c = -0.06) and for pCO_2 (d = 0.25) are in qualitative agreement with data by Lauvstad for Ni point anodes. An activation energy of $E_{act,an}$ = 1.23 eV is reported. The dominance of H_2 oxidation during the oxidation of CO-H_2 has been shown in a study by Weber et al. for Ni/8YSZ cermet anodes [50].

2.2 Modeling the Reaction Mechanisms

The major unsolved question in research on SOFC anodes is the reaction mechanism of the electrochemical oxidation of the fuel gas (i.e. H_2, CO and reformate gases). Here, detailed modeling studies are required in addition to experimental results from electrochemical characterization.

Depending on the aim of the study, modeling of the SOFC is performed on different levels. Sorted by decreasing scale, these levels include the system level, stack level, cell level, electrode level, surface level and finally the molecular and interphase level, for a detailed description see [15]. Following this definition, the surface level describes the elementary kinetic chemistry and surface diffusion, yielding information on the reaction pathways and rate limiting steps. It is also commonly denoted elementary kinetic modeling.

In contrast to global modeling e.g. on the cell and electrode level, where global kinetic expressions such as the Butler-Volmer equation are applied, the elementary kinetic modeling approach consists of the description of the individual physical or chemical step reactions of surface and gas phase transport, heterogeneous chemistry (i.e. adsorption / desorption) and charge transfer by mass action laws. This allows the modeling in non-equilibrium conditions and the calculation of parallel charge transfer pathways [4, 54]. However, it also leads to a complex framework including numerous parameters such as thermodynamic data of all involved species and diffusion coefficients. The determination of these parameters is often difficult due to a lack of experimental data in the temperature range of interest or considerable deviations between results of different studies [16, 55].

For the parameter adaptation as well as the experimental validation of the elementary kinetic models, simplified geometries such as model anodes are used preferentially. Several modeling studies have been performed so far both for the H_2-H_2O and the CO-CO_2 atmosphere, an overview is given in the following sections.

2.2.1 H_2-H_2O Atmospheres

For the modeling of the H_2-H_2O atmosphere, several elements (e.g. H_2, H_2O, O^{2-}) in different states (e.g. in the gas phase or surface adsorbed on Ni or YSZ) need to be considered and are typically denoted species. For example, Vogler et al. have considered a total of 12 species, interacting by 8 surface reactions and 7 charge transfer reactions [16]. In order to reduce the modeling complexity, most studies have made certain simplifications e.g. reduced the dimensions of the system using simplified geometries, disregarded some contributions or applied single step mechanisms exclusively.

For the validation of the modeling studies, the problem of the discrepancies between different experimental data sets persists, as already discussed in section 2.1.1. Due to the different absolute LSR values and the discrepancies in parameter dependencies, it is not possible to cover all experimental studies with a single model.

table 2.3: Comparison of elementary kinetic modeling studies reported in literature for the H_2-H_2O atmosphere [13, 16, 54-57] with respect to the type of model, the considered processes ('CT' indicates the charge transfer mechanism), the experimental data set used for validation and conclusions on the rate determining reaction step. Additionally, some simplifications taken in the respective model are given.

Study	Ref.	Type	Processes	Exp data	Result	Comments
Bieberle	[55]	State Space Model Single step 0D	1 CT: O spillover + heterogen. chem.	Bieberle [6]	No definite conclusion on rate limiting step possible	Ni surface active, YSZ surface inactive, no surface & gas phase transport, scaling of kinetic parameters
Bessler	[56]	Elementary mass action kinetics Single step 1D	5 CT: O spillover, OH spillover, H spillovers, H interstitial	Mizusaki [8] de Boer [7] Bieberle [6]	Mizusaki: H spillover or H interstitial De Boer: no clear trend Bieberle: H spillover to OH	Only charge transfer, no surface & gas phase transport, no heterogen. chem. exchange current density scaled to exp. data series
Vogler	[16]	Elementary mass action kinetics Single & two step 1D	7 CT: O spillover, H spillover, OH spillover + surface transport + heterogen. chem.	Bieberle [6]	Two H spillovers (to O and OH) yield best agreement	Scaling of rate coefficients, adjustment of diffusion coefficients
Vogler	[57]	Elementary mass action kinetics Single step 2D	1 CT: interstitial + surface & bulk transport + heterogen. chem.	Bieberle [6]	Interstitial H transfer feasible, but surface spillover preferred	Scaling of rate coefficients, adjustment of diffusion coefficients
Bessler	[13]	Elementary mass action kinetics Single & two step 1D	2 CT: H spillover, + surface transport + heterogen. chem.	Utz [58]	Change in rate determining step for different temperature (co-limitation)	Scaling of rate coefficients, adjustment of diffusion coefficients
Goodwin	[54]	Elementary mass action kinetics Single & two step 1D	4 CT: O spillover, H spillover + surface transport + heterogen. chem.	Mizusaki [8]	Dual channel H spillover represents data well	Scaling of rate coefficients, adjustment of symmetry factor

Therefore, most studies limit the model validation to a single experimental data set, which considerably compromises its significance. Furthermore, different studies have focused on different aspects of the electrochemical characterization and no complete data set exists: data for cathodic polarization is reported exclusively by Mizusaki, while a large temperature variation is only performed by Bieberle. In an early study, Bessler and coworkers have compared modeling results to all available experimental data sets, but applied other simplifications that affect the modeling result: only charge transfer was considered and – due to the unknown preexponential factors and activation energies - the simulated exchange current densities were scaled to experimental data [56]. Nevertheless, this work emphasizes the difficulty to draw conclusions based on an inconsistent experimental data set. Later on, Bessler and coworkers only used data by Bieberle for the validation [16, 57], whereas Goodwin et al. validate their model with data by Mizusaki [54]. Thus, a direct comparison of the two modeling approaches is not possible.

An overview of modeling studies with patterned Ni anode data is given in table 2.3. Despite the differences in (i) the modeling approach, (ii) the number of included charge transfer mechanisms, (iii) the simplifications regarding heterogeneous chemistry, and in (iv) scaling of results to experimental data, the general consensus is that charge transfer is the rate limiting step and most probably proceeds as H spillover. However, the actual procedure is not clear: whereas Bessler and Vogler suggest a temperature dependent co-limitation of H spillover to O and OH [13, 16], Goodwin et al. are in favor of a dual channel H spillover, where the mechanism of H spillover to OH appears twice with different rate constants [54].

Only one study considers interstitial charge transfer that is proposed in the context of a blocking of the TPB, see section 2.3 [4]. Vogler was able to reproduce the experimentally determined parameter dependencies of pH_2, pH_2O and T of the study by Bieberle, but it is questionable, whether the measured exchange current density can be obtained with the diffusion coefficients for interstitial H in the YSZ bulk [57]. However, a certain amount of the exchange current density may be supplied by this mechanism in parallel to a surface spillover mechanism [59].

Further modeling studies on the electrochemistry exist for cermet anodes [39, 60], but will not be treated in detail in this work. A comprehensive review is given by Adler and Bessler in [4].

2.2.2 CO-CO_2 Atmospheres

Only few modeling studies have been carried out with CO-containing fuels so far [14, 49, 61, 62]. In an early study, Etsell et al. [62] characterized porous Pt anodes in CO-CO_2 atmosphere and found that two species are involved in the reaction. However, no definite conclusion was possible whether – besides gaseous CO_2 – gaseous or adsorbed CO was involved. A combined modeling and experimental study has been performed by Lauvstad et al. including four different mechanisms for charge transfer [14, 49]. The application to experimental data from Ni point anode characterization indicates a reaction mechanism involving at least two adsorbed intermediates that were assumed to be O_{ad} and CO_{ad}.

Recently, Yurkiv et al. performed a modeling study on the CO-CO_2 reaction mechanism combined with Temperature Programmed Desorption (TPD) measurements to determine the kinetics of adsorption/desorption [61]. For the charge transfer, five different mechanisms have been modeled. Best agreement to experimental data from Lauvstad et al. [49] was obtained for a two step charge transfer consisting of a transfer of one electron on the YSZ surface and a consecutive O spillover from YSZ to the Ni surface (see equations 1.10 and 1.11).

A comparison of the modeling studies is given in table 2.4.

table 2.4: Comparison of elementary kinetic modeling studies reported in literature for the CO-CO_2 atmosphere [14, 49, 61] with respect to the type of model, the considered processes ('CT' indicates the charge transfer mechanism), the experimental data set used for validation and conclusions on the rate determining reaction step. Additionally, some comments to the respective model are given.

Study	Ref.	Type	Processes	Exp data	Result	Comments
Lauvstad	[49]	Elementary mass action kinetics Single & two step 0D	4 CT: O spillover CO spillover + heterogen. chem. + surface transport	Lauvstad [49]	Rate determining step with 2 adsorbed species (O_{ad}, CO_{ad})	Calculation of steady state relationships CT at TPB or electrode-electrolyte interface
Yurkiv	[61]	Elementary mass action kinetics Single & two step 1D	5 CT: O spillover + heterogen. chem. + surface transport	Lauvstad [49]	Two step O spillover represents data well	Scaling of rate coefficients, adjustment of diffusion coefficients

2.3 Effect of Impurities

Impurities at the surface of electrode or electrolyte, and especially at the TPB, are considered as barriers for the charge transfer reaction. They are suspected to influence the charge transfer mechanism, e.g. in favor of interstitial charge transfer [4, 63]. As a consequence, the avoidance of impurities as well as the knowledge of the impurity amount is of special interest in the analysis of the elementary kinetics of SOFC anodes.

A low amount of impurities is always present in the raw material of the electrolyte, since their complete removal is very expensive. In some cases, impurities are even added intentionally in order to improve the sintering behavior [64]. Additionally, impurities can be introduced by the preparation method, during sintering in an oven with inappropriate crucible or during the surface cleaning process. Due to the high surface energy of unrelaxed YSZ and the high affinity of SiO_2 to the surface, a strong driving force for impurity segregation to the surface exists [18]. The segregation and accumulation of impurities at the surface leads to the formation of a glassy alkali silicate phase [65].

As the impurities on the surface originate mainly from the bulk material, the surface to volume ratio is a decisive parameter [66]. In this matter, model anodes are disadvantageous compared to micro- or nano-sized cermet structures and therefore require special focus on (i) the avoidance of impurities and (ii) the determination of the impurity amount.

So far, no impurity analysis has been performed for patterned Ni anodes. However, a considerable amount of impurity analyses has been performed for Ni point anodes by Mogensen and coworkers [18, 21-23, 26, 34, 63, 65-73]. In addition to the electrochemical characterization of Ni point anodes, the impurity content on the surface of the electrodes has been determined by various surface analysis tools, including Energy Dispersive X-ray Diffraction (EDX), X-ray Photoelectron Spectrometry (XPS), and Time-Of-Flight Secondary Ion Mass Spectroscopy (TOF-SIMS). The analyses include comparisons of different purity levels of the Ni electrode [23, 65, 68, 73] as well as of different electrolytes (YSZ and ScYSZ) [26, 70, 71]. Furthermore, studies have been performed at a lower temperature of 500 °C in order to reduce impurity segregation [67].

A difference in the initial behavior during the first 50 h of thermal exposure at 1000 °C was observed for samples with two different purity levels of the Ni electrode (99.8 % and 99.995 %): while for the purer electrode the LSR decreases, which is attributed to an increasing l_{TPB}, the LSR of the less pure material increases due to a build-up of a high impurity ridge. Even though considerable sample to sample scatter superimposes the data, the observed trend is that the LSR value of the impure electrode is up to an order of magnitude higher than for the pure electrode [23]. At the same time, both observed impurity features – the impurity ridge and a hill-and-valley structure at the electrode-electrolyte interface – are more pronounced for the less pure Ni electrode.

For two different purity levels of ScYSZ electrolyte substrate, a difference in absolute LSR values has been observed with the higher value for the less pure electrolyte. The only difference in impurity content is observed for the electrode-electrolyte interface, whereas the electrolyte surface outside the contact area as well as the impurity ridge at the TPB exhibit no perceivable differences in height and composition. Furthermore, a lateral difference in impurity concentration has been observed, distinguishing between a core zone with high impurity content and a rim zone with low impurity content. The zone with reduced impurity content is believed to be caused by metallic creep during the initial exposure to temperatures of 850 – 1000 °C [71].

From these two observations it is concluded that the reaction mechanism takes place at the electrode-electrolyte interface, more precisely in the purer rim zone. In this case, a proton transport across the electrode-electrolyte interface is the most probable step [26, 71].

Further studies have investigated the effect of impurities: Hughes et al. found that the surface of a YSZ single crystal is fully saturated with SiO_2 after 300 min at 900 °C [74]. Calculations based on thermodynamic equilibrium segregation show that impurity levels below 10 ppm are necessary to avoid a full coverage of the surface by impurities [75]. However, the surface to volume ratio of the electrolyte is also a decisive point. Furthermore, the origin of Si mainly from the electrolyte is validated by a study by Høgh with an electrolyte wrapped up in a Pt foil [67]. De Ridder et al. attempted a depletion of the bulk impurities by steps of sintering and consecutive cleaning of the surface using Ar sputtering [75]. However, within 13 steps no considerable improvement was achieved.

Based on the above findings, a model for the different effects observed during electrochemical characterization of Ni point anodes has been setup by Mogensen and coworkers that includes the presence of impurities:

(i) Initial behavior: The assumption of a reaction by interstitial transport through the bulk (and not across the TPB barrier) is supported by the initial behavior of the point anode. During creep of the Ni electrode, a rim zone of high purity is forming at the electrode-electrolyte interface, explaining the decrease in LSR described by Schmidt et al. [71].

(ii) Effect of water: The glassy impurity layer that acts as barrier for the charge transfer reaction changes its electrical properties upon uptake of water from the atmosphere. Whereas the impurity ridge is almost entirely electrically insulating in dry atmosphere, it is getting considerably proton conducting for increasing water vapor content, leading to a decrease of LSR [66]. As further explanation, the formation of highly volatile $Si(OH)_4$ and thus a decrease of impurity content with increasing water vapor content is suggested [63].

(iii) Activation during polarization: The short term activation observed for anodic and cathodic polarization is assumed to be caused by a change in the electrical field at the interface and a consecutive redistribution of the impurities at the electrode-electrolyte interface yielding an unblocked area. This is supported by the fact that the activation for pure electrodes is less pronounced than for impure electrodes [65].

(iv) Current oscillation: The observed current oscillation during polarization is attributed to a short-term passivation of the Ni surface by $Ni(OH)_2$ and NiO formation, or a coupling of both [71, 72].

Based on these arguments, it is concluded that the contradictions found in earlier literature on model anodes may be due to different impurity levels and different impurity compositions [63]. For instance, a strong difference in resistance of soda-silicate glasses as a function of material composition has been shown by Seddon et al. [76].

While a detailed analysis of Ni point anodes has been performed by the comparison of absolute LSR values of samples with different impurity levels, the parameter dependencies on the partial pressures of H_2 and H_2O as well as on temperature have not been determined. Hence, no definite conclusion on a change in reaction mechanism as a function of impurity content is possible.

2.4 Conclusions

The strong deviations in existing literature on the electrochemical characterization of patterned Ni anodes in H_2-H_2O atmosphere emphasize the need of a new data set with parameter variations of pH_2, pH_2O, T and η over a large range. Additionally, explanations for the differences are required. In this work, the characteristics of patterned Ni anodes including stability and reproducibility as well as an impurity analysis will be addressed in chapters 1 and 4. A new, comprehensive data set of the electrochemical characterization of patterned Ni anode in H_2-H_2O is determined in chapter 5.

So far, no data set of the electrochemical characterization of patterned Ni anodes in CO-CO_2 atmosphere has been reported that includes the parameter dependencies on pCO, pCO_2 and T. However, such a data set is strongly needed, e.g. for the validation of modeling studies. In this work, a comprehensive data set for the characterization in CO-CO_2 atmosphere is established and the absolute LSR values and parameter dependencies are compared to the results obtained in H_2-H_2O atmosphere.

Furthermore, more studies on the determination of the parameters used in the elementary kinetic modeling studies are needed for the temperature range relevant for SOFCs.

3 Characteristics of Patterned Ni Anodes

In this chapter, the general characteristics of patterned Ni anodes as they are used in this work are described. The fabrication and stability of the samples is addressed in section 3.1 and the measurement setup is explained in section 3.2. In section 3.3, the equivalent circuit for the interpretation of the impedance spectrum is introduced. The results of a detailed impurity analysis of the characterized patterned anodes are given in section 3.4.

3.1 Preparation of Patterned Ni Anodes

3.1.1 Fabrication Procedure

Prior to the characterization of the patterned Ni anodes, a comparative analysis of different fabrication procedures has been performed with emphasis on reproducible sample properties, good edge quality and feasible pattern design [77]. The results reported in this work have all been obtained by the same fabrication procedure, which is specified in the following.

Patterned Ni anodes were fabricated on 8.5 mol% polycrystalline Y_2O_3 –stabilized ZrO_2 (8YSZ) substrates (Itochu Ceratec Corp., Japan) by a standard photolithography process [78].

For structuring, the substrates were covered with a two layer resist system, ensuring an undercut pattern for easy lift-off. Both bottom-resist layer (Allresist AR-P 5480) as well as photoresist (Allresist AR-P 3540) were applied employing spin coating with 3000 and 6500 rpm for 90 sec respectively. After bottom-resist coating, the substrates were prebaked at 150 °C for 5 min, whereas prebaking of the photoresist layer was carried out at 100 °C for 2 min. Structuring of the substrates was performed using a SUSS MJB300/400 mask aligner working with UV-light in the 300 - 400 nm range. The exposure time was adapted to the mercury vapor lamp yield and the required photoresist dose (50 - 100 $mJ \cdot cm^{-2}$). After structuring of the patterns, 800 nm thick Ni layers were deposited by magnetron sputtering (Von Ardenne CS 730S, 60 W, no bias, purity of Ni: 99.98 %), followed by the lift-off (Allresist AR300-72) of the unexposed regions of the patterns. This fabrication procedure enabled reliable and reproducible manufacturing of patterned Ni anodes. The adhesion of the Ni pattern on the substrate proved to be good: no removal of Ni was observed after a scotch test.

The purity of the raw materials – YSZ substrate and Ni sputter target – is given in table 3.1 and table 3.2.

table 3.1: Purity of the Ni target: impurity species and their amount as stated by the target supplier (FHR Anlagenbau GmbH, Germany).

Species	Concentration [w. %]
Ni	99.98
Zn	0.0001
Fe	0.0014
Co	0.0029
Cu	0.0019
Pb	0.0040

table 3.2: Purity of the YSZ substrate: impurity species and their amount as stated by the supplier (Itochu Ceratec Corp., Japan).

Species	Concentration [w. %]
Y_2O_3	13.79
Al_2O_3	0.074
SiO_2	0.032
Fe_2O_3	0.003
MgO	0.001
Na_2O	0.035
Hf_2O	1.6

3.1.2 Pattern Design

In this work, two different pattern designs have been prepared, see figure 3.1. In both designs, the Ni pattern is positioned in the center of the 20 x 20 mm^2 YSZ substrate. Pattern type #1 with a size of 10 x 10 mm^2 is composed of parallel stripes and two perpendicular stripes at the ends that ensure contacting of all stripes. Pattern type #2 with a size of 10.25 x 10.25 mm^2 is composed of an area of 9.25 x 9.25 mm^2 with parallel Ni stripes and a 500 µm wide frame that allows contacting with a special contact block. In both designs, different TPB lengths between 1 to 8 m·cm^{-2} have been realized by a variation of stripe width and spacing, see sample list in section 9.4.

The reproducible sample quality was affirmed by SEM and light-microscopy analyses of the samples before the electrochemical characterization. SEM images at two different magnifications are shown in figure 3.2.

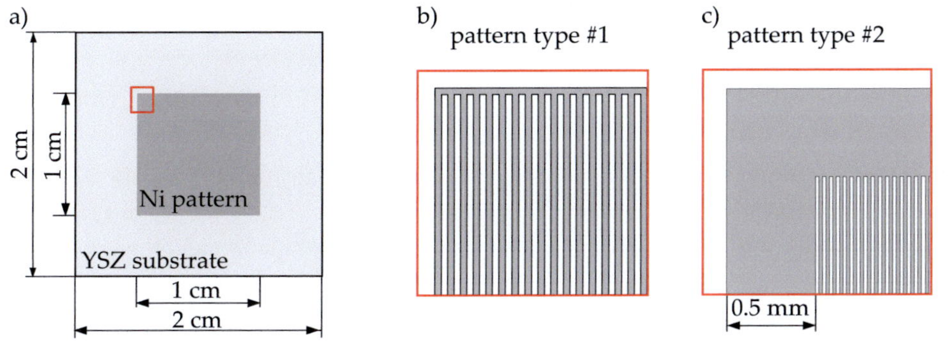

figure 3.1: Scheme of pattern design for the patterned Ni anodes prepared in this work: a) the Ni pattern of size 10 x 10 mm² (type #1) or 10.25 x 10.25 mm² (type #2) is positioned in the center of the 20 x 20 mm² YSZ substrate, b) pattern type #1 consists of parallel stripes, two perpendicular stripes at the ends ensure a contacting of all stripes, and c) pattern type #2 consists of parallel stripes within a 500 μm wide frame, which allows contacting with a special contact block. For information on stripe width and spacing see the list of samples in section 9.4.

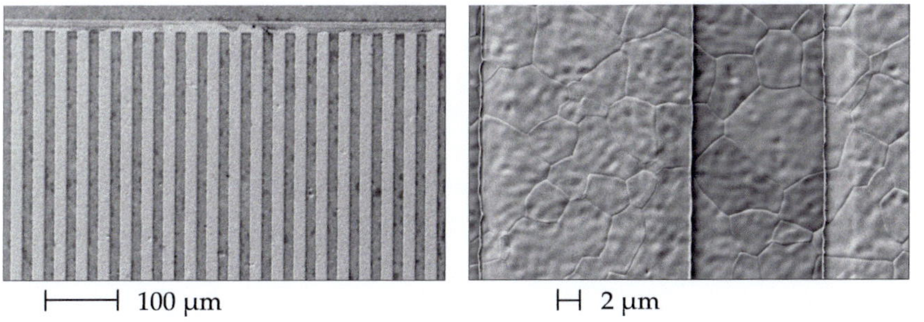

figure 3.2: SEM images at two different magnifications of a prepared patterned Ni anode with pattern design #1: 15 μm stripe width / spacing and 15 μm wide perpendicular stripe at the end, layer thickness: 800 nm.

3.1.3 Stability Analysis

The stability of Ni thin films on YSZ substrates has been addressed in a detailed study and the findings have been published in [51]. In the following, the key results are summarized briefly.

The analysis of samples with different Ni layer thickness between 50 nm to 1.2 μm for a variation of temperature and gas composition revealed that the stability of a Ni thin film strongly depends on (i) exposure temperature, (ii) exposure time, (iii) layer thickness, and (iv) water vapor content of the anode gas. Thin Ni layers exhibit considerable grain growth followed by hole formation in the continuous film and a separation into Ni islands with increasing temperatures. An increase in water vapor content leads to an acceleration of this degradation process. Thicker Ni layers proved to be more stable at elevated temperature, yet the average grain size also increases with Ni layer thickness and leads to a roughening of the formerly straight and sharp edge. However, the exact knowledge of the TPB length is required for the characterization of patterned Ni anodes. It is therefore important to keep the Ni layer thickness as thin as possible to reduce the increase in l_{TPB}, and also the time of stabilization that is required until the morphological changes (i.e. grain growth) have attained equilibrium. Additionally, thinner Ni layers reduce the risk of delamination of the Ni film due to internal stress, and the smaller grain size allows the fabrication of narrow stripes with a minimum stripe width of 5 μm at a layer thickness of about 1 μm, see also [9].

A Ni layer thickness of 800 nm was determined to assure the stability of the patterned Ni anode during the electrochemical characterization at temperatures of 700 - 900 °C in H_2-H_2O atmosphere for water vapor contents up to β_{rel} = 90 %. Therefore, all patterned Ni anodes characterized in this work were prepared with a Ni layer thickness of 800 nm. SEM analysis after the electrochemical characterization of the patterned Ni anodes in CO-CO_2 atmosphere for 700 °C ≤ T ≤ 800 °C confirmed the stability in this atmosphere as well.

The above mentioned morphological changes lead to an initial activation of LSR through an increase in l_{TPB} and in in-plane conductivity. After about 20 h, this activation reaches equilibrium with no further grain growth during the subsequent electrochemical characterization. Therefore, an initial heat treatment with fixed temperature, gas composition and duration was established in order to achieve reproducibility of the Ni pattern. For more details see section 4.1.

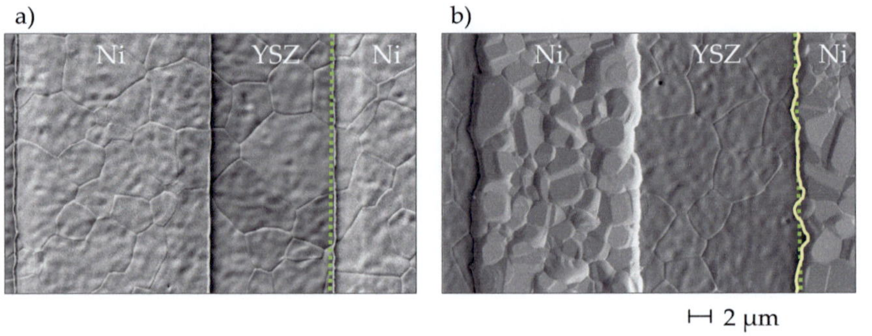

figure 3.3: SEM images of patterned anodes: a) as prepared with straight and sharp edges. The dotted line indicates $l_{TPB,th}$, and b) after electrochemical characterization at 700 °C ≤ T ≤ 800 °C, $6.9\cdot10^3$ Pa ≤ pH_2O ≤ $8.3\cdot10^4$ Pa, $1.8\cdot10^4$ Pa ≤ pH_2 ≤ $8.7\cdot10^4$ Pa, 13 % ≤ β_{rel} ≤ 82 %, balance N_2, exposure time = 210 h. The solid line indicates the increase in l_{TPB} during characterization compared to the ideal $l_{TPB,th}$ (dotted line).

The increase in l_{TPB} due to the grain growth needs to be considered for the interpretation of results from electrochemical characterization. Therefore, the actual l_{TPB} ($l_{TPB,corr}$) was determined based on SEM images using the image processing tool SPIP [79]. A correction factor of τ_{corr} = 1.3 was obtained as quotient of $l_{TPB,corr}$ and the theoretical l_{TPB} ($l_{TPB,th}$), see figure 3.3. The average value determined for different positions on one sample as well as for several samples is almost constant and is therefore taken as a standard value. The same value was observed after electrochemical characterization in H_2-H_2O and CO-CO_2 atmosphere, which affirms that no further grain growth occurs in the Ni thin film after the initial heat treatment.

3.2 Measurement Setup

Cell Setup

For the characterization in this work, a pseudo-symmetric cell composed of two different anodes has been chosen: the patterned Ni anode is the working electrode, while a screen-printed Ni/8YSZ cermet anode is applied as counter electrode. Herein, the contribution of the counter electrode is known and even negligibly small compared to the contribution of the patterned anode. This cell setup differs intentionally from previous studies in literature that characterized the patterned Ni anode against a reference electrode [6-9] in order to avoid the inaccuracies related to a characterization with reference electrodes. Adler et al. showed that even for perfectly aligned electrodes, inherent distortions of the impedance such as frequency dispersion and inductive artifacts occur [80].

The Ni/8YSZ counter electrode was screen-printed following the procedure described by Sonn [35]. In the same contribution, the electrochemical characteristics of this type of electrode have been identified. This counter electrode was applied to the substrate prior to the preparation of the patterned Ni anode, since the fabrication steps of sintering at 1300 °C and NiO reduction at 850 °C require the highest temperatures.

Contacting

According to the pattern design, two different contacting methods have been realized in this work, see figure 3.4. The samples with pattern design #1 are contacted on both sides by a fine Ni-mesh (3487 wires·cm^{-2}, 0.065 mm wire thickness); the counter electrode is positioned on the Al_2O_3 sample holder. An additional weight of 18 g on top of the pattered anode contact mesh ensures the electrical contact for both electrodes. The samples with pattern design #2 are contacted by a special contact block that is positioned on two sides of the frame perpendicular to the stripe pattern, see figure 3.4b. A fine Ni mesh on the contact surface ensures uniform contacting. The counter electrode is contacted by a combination of a fine and coarse Ni mesh on a Ni sample holder with flow field. Compared to the first setup, these modifications have reduced the influence of gas diffusion on overall electrode performance through direct gas flow to the patterned Ni anode and improved gas supply to the counter electrode. At the same time, the contribution of charge transfer is unaffected by these modifications (see section 9.1).

Measurement Setup

The sample holder is placed in a gastight Al_2O_3 chamber of a horizontal tube furnace with single chamber setup, see figure 3.5. The electrochemical characterization is performed at ambient pressure with variations of pH_2, pH_2O or pCO, pCO_2 and pN_2 in between 0 and 100 % at a gas flow of 100 – 300 ml·min^{-1}. The independent variation of pH_2, pH_2O, pCO and pCO_2 respectively is realized by the addition of inert N_2 as balance gas. Herein, gas composition and gas flow are controlled by a mass flow controller (MFC) battery, denoted gas mixing battery. The employed measurement setup was developed at the Institute of Materials for Electrical and Electronic

Engineering (IWE) and is optimized for the characterization of patterned Ni anodes in very dry atmospheres.

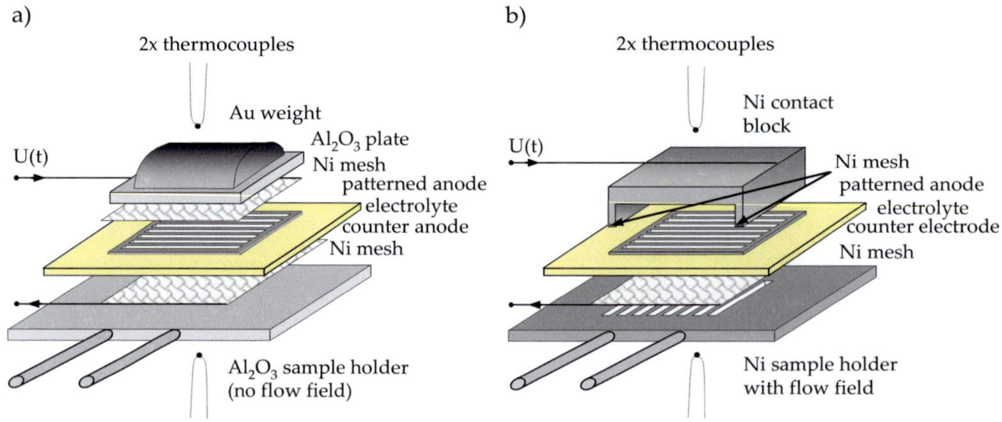

figure 3.4: Scheme of the measurement setup for the electrochemical characterization of patterned Ni anodes: a) contacting of the patterned anode with Ni mesh on the entire electrode surface. An Al_2O_3 plate and an additional Au weight ensure good electrical contact. The counter electrode is contacted by a combination of fine and coarse Ni mesh on top of the Al_2O_3 sample holder without flow fiel, b) contacting of the patterned anode with a special Ni contact block only on two opposite sides of the frame. A fine Ni mesh on the contact area ensures electrical contact. The counter electrode is contacted by a fine Ni mesh on top of a Ni sample holder with flow field.

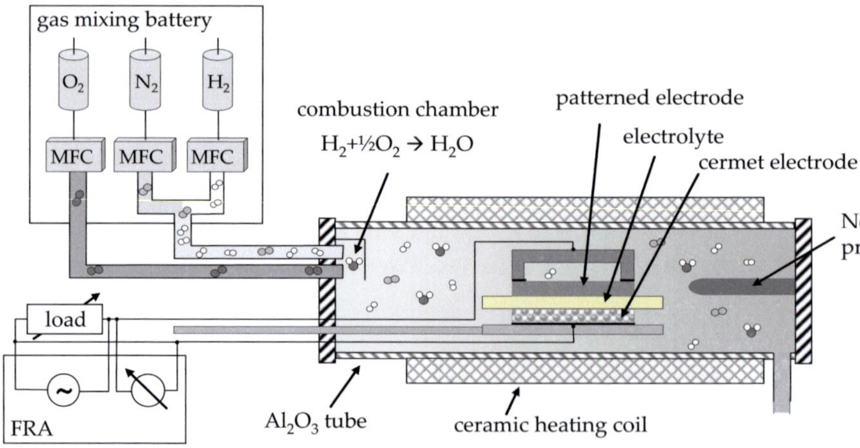

figure 3.5: Scheme of the measurement setup for the characterization of patterned Ni anodes in a horizontal tube furnace. The gas composition in the setup is precisely adjusted by the gas mixing battery. The humidification of the gas is realized by an upstream combustion chamber and monitored by a Nernst probe.

Gas composition and control

In this work, patterned Ni anodes have been characterized in H_2-H_2O as well as in CO-CO_2 atmosphere. The parameter dependencies of LSR_{ct} on pH_2, pH_2O, pCO and pCO_2 were determined by individual variation of the respective partial pressure by the addition of N_2.

The humidification of the gas was realized by feeding oxygen and hydrogen into an upstream combustion chamber. With this method it is possible to obtain very stable gas conditions even at high water vapor contents. A Nernst probe located near the sample monitors the partial pressures. This allows to determine the actual gas composition instead of the theoretical calculation based on the flow rates of the gases. Thus, possible leakages in the setup are taken into account. As the set-up has been designed to ensure maximum gas tightness, very low leakage rates were achieved (< 0.02 %).

For the selection of gas variation range, two restrictions need to be considered: (i) oxidation of Ni at elevated oxygen partial pressures following the oxidation reaction (equation 3.1), and (ii) solid carbon formation at high concentrations of CO following the Boudouard reaction (equation 3.2) with subsequent metal dusting of Ni [81].

$$\mathrm{Ni} + \frac{1}{2}\mathrm{O_2} \rightleftharpoons \mathrm{NiO} \qquad 3.1$$

$$2\mathrm{CO} \rightleftharpoons \mathrm{C} + \mathrm{CO_2} \qquad 3.2$$

Both limits have been calculated using the thermodynamic database MALT [82] and the range of parameter variation of pH_2-pH_2O and pCO-pCO_2 that avoids these effects is indicated for different temperatures in figure 3.6. At 800 °C, the critical pO_2 for oxidation of Ni is $1.2 \cdot 10^{-9}$ Pa.

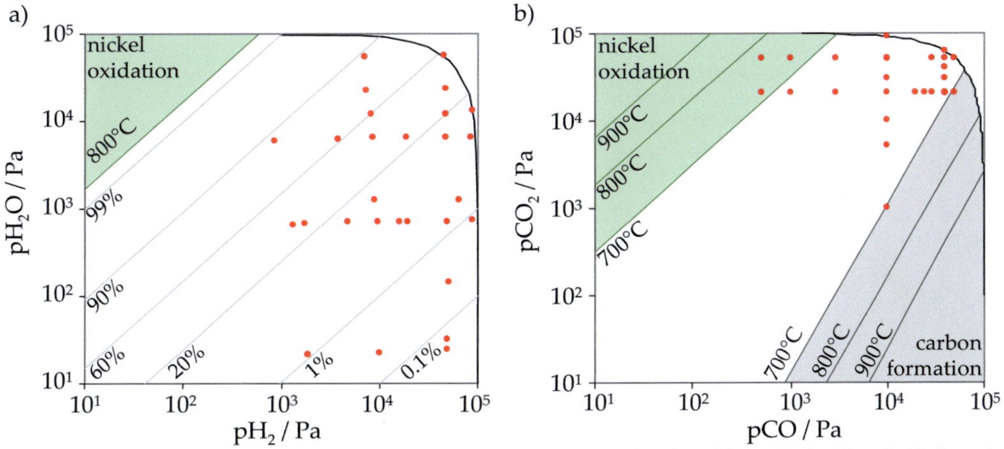

figure 3.6: Summary of parameter variation range of gas composition realized in this work: a) pH_2 and pH_2O variation. The limit of Ni oxidation at 800 °C, as well as the respective relative water vapor content β_{rel} are indicate, and. b) pCO and pCO_2 variation. The critical regions of carbon formation and of Ni oxidation are indicated for temperatures of 700, 800 and 900 °C. The red dots indicate parameter conditions that have been realized for the electrochemical characterization, the black line indicates the limit of atmospheric pressure.

In figure 3.6, the realized gas variations for both atmospheres are indicated. Compared to previous literature on patterned Ni anodes (see figure 2.1), wide ranges of the partial pressures are realized with $8 \cdot 10^2$ Pa $\leq pH_2 \leq 9 \cdot 10^4$ Pa, $2 \cdot 10^1$ Pa $\leq pH_2O \leq 6 \cdot 10^4$ Pa, as well as $4.0 \cdot 10^2$ Pa $\leq pCO \leq 5.1 \cdot 10^4$ Pa, $9.5 \cdot 10^2$ Pa $\leq pCO_2 \leq 9.2 \cdot 10^4$ Pa. Note that in CO-CO$_2$ atmosphere, the characterization was performed in all gas compositions at 800 °C, while for the characterization at 700 °C the critical gas compositions have been omitted.

In this work, the gas compositions are, in addition to the indication of the gas partial pressures, described by their ratio of reaction product to reactant. For H_2-H_2O atmosphere, the ratio is denoted relative water vapor content β_{rel} and calculated according to equation 3.3. In CO-CO_2 atmosphere, the ratio is indicated by $\beta_{rel,CO/CO2}$ and calculated according to equation 3.4. The variation of gas compositions in figure 3.6 corresponds to large variations of 0.02 % ≤ β_{rel} ≤ 88 % and 8 % ≤ $\beta_{rel,CO/CO2}$ ≤ 99 %. The characterization for such a large parameter range is important to identify possible changes in the reaction mechanism as a function of gas composition.

$$\beta_{rel} = \frac{pH_2O}{pH_2O + pH_2} \quad 3.3$$

$$\beta_{rel,CO/CO2} = \frac{pCO_2}{pCO_2 + pCO} \quad 3.4$$

3.3 Equivalent Circuit Development

For the interpretation of impedance spectra, an equivalent circuit has been developed that is appropriate for the interpretation of all impedance spectra in this work. This model was first established for H_2-H_2O atmosphere and proved to be applicable in CO-CO_2 atmosphere as well.

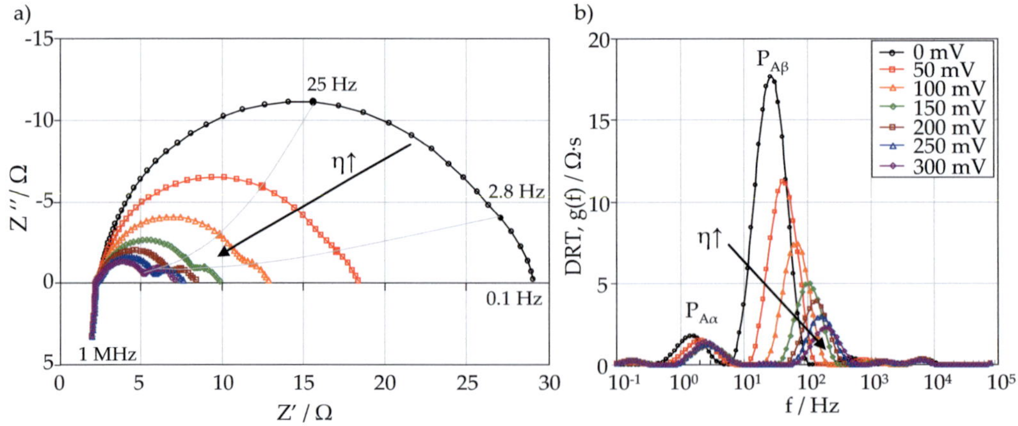

figure 3.7: a) Impedance spectra plotted for a variation of polarization voltage between 0 mV ≤ η ≤ 300 mV at constant conditions (800 °C, $pH_2 = 8.3\cdot10^3$ Pa, $pH_2O = 6.7\cdot10^3$ Pa, balance N_2, β_{rel} = 45 %, sample: NiYSZ_146), and b) corresponding DRT spectra. A clear separation of the two processes $P_{A\alpha}$ and $P_{A\beta}$ is observable for increasing polarization voltage.

In figure 3.7, examples of impedance spectra for patterned Ni anodes are given together with the corresponding result of the Distribution of Relaxation Times (DRT) calculation [83]. This methodological approach allows to clearly separate the time constants of two low frequency processes, particularly at an elevated polarization voltage η. Altogether, three different processes are distinguished in patterned Ni anodes ($P_{A\alpha}$, $P_{A\beta}$, $P_{A\gamma}$). The respective physical origin is identified by a systematic parameter variation of gas composition, temperature and polarization voltage. This procedure has recently been presented for highly efficient Ni/8YSZ cermet anodes in [33]:

(i) The low frequency process $P_{A\alpha}$ at a relaxation frequency of 1 to 6 Hz and its dependency on temperature and gas composition is shown in figure 3.8. The temperature activation with $E_{act} = 0.03$ eV is negligible. The dependency on gas composition is determined for binary H_2-H_2O mixtures without addition of N_2. The proportionality of the resistance $R_{A\alpha}$ is expressed in equation 3.5, which has been introduced by Primdahl [84]. Both parameter dependencies clearly assign $P_{A\alpha}$ to gas diffusion. The small influence of polarization voltage on this process is explained by the small currents (< 20 mA at $\eta = 300$ mV) and the small amounts of gas converted at the TPB of patterned Ni anodes.

$$R_{A\alpha} \propto \left(\frac{1}{x_{H_2O}} + \frac{1}{1 - x_{H_2O}} \right) \quad (3.5)$$

(ii) Process $P_{A\beta}$ shows strong temperature activation and depends on gas composition and polarization voltage. This behavior is attributed to the charge transfer process in the patterned Ni anode. The relaxation frequency ranges from 0.1 to 100 Hz, the area specific resistance (ASR) is by far the largest with ASR = 5.5 to $3.4 \cdot 10^3$ $\Omega \cdot cm^2$.

(iii) The high frequency process $P_{A\gamma}$ shows strong dependencies on temperature and gas composition. The extracted values (relaxation frequency $5 \cdot 10^3$ - 10^4 Hz and a very small ASR < 1 $\Omega \cdot cm^2$) are characteristical for highly efficient Ni/8YSZ cermet anodes [35], which have been applied as counter electrode in this work.

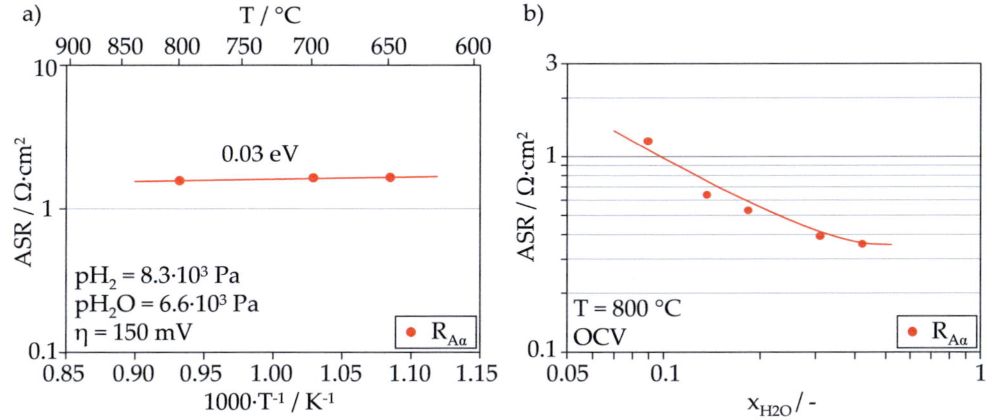

figure 3.8: Parameter dependencies of $P_{A\alpha}$: a) dependency of $R_{A\alpha}$ on temperature together with the activation energy $E_{act} = 0.03$ eV ($pH_2 = 8.3 \cdot 10^3$ Pa, $pH_2O = 6.6 \cdot 10^3$ Pa, balance N_2, $\beta_{rel} = 44$ %, $\eta = 150$ mV, sample: NiYSZ_140), and b) dependency of $R_{A\alpha}$ on variation of the molar fraction of H_2O together with the fit according to equation 3.5. Only binary gas mixtures are studied, thus x_{H2O} equals pH_2O ($T = 800$ °C, $\eta = 0$ mV, sample: NiYSZ_148).

A summary of the resistance R, the relaxation frequency f_{rel} and the exponent n of the processes $P_{A\alpha}$, $P_{A\beta}$ and $P_{A\gamma}$ is given in table 3.3. The relaxation frequency is calculated by $f_{rel} = (2\pi\cdot\tau)^{-1}$. The exponent n is extracted from $Z_{RQ} = R/(1+(j\omega)^n RQ)$, where n indicates the deviation from an ideal RC behavior with n = 1 [85]. The patterned Ni anode shows only a small deviation with $n_{A\beta} = 0.9 - 0.95$, whereas the Ni/8YSZ cermet counter anode shows a distinct deviation with $n_{A\gamma} = 0.7 - 0.78$. The gas diffusion process $P_{A\alpha}$ is almost ideal with $n_{A\alpha} = 0.95$. The resulting equivalent circuit model is composed of three RQ elements ($RQ_{A\alpha}$, $RQ_{A\beta}$, $RQ_{A\gamma}$), see figure 3.9.

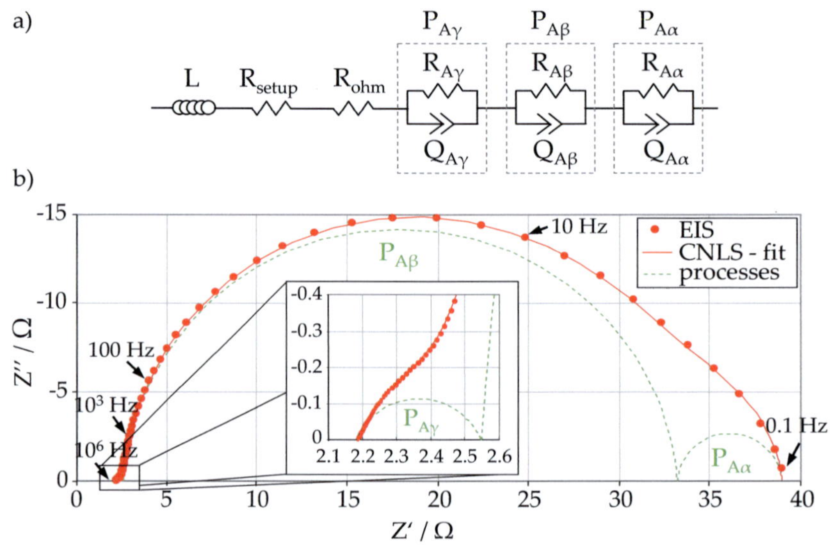

figure 3.9: a) Equivalent circuit model with three RQ-elements representing contributions of gas diffusion ($P_{A\alpha}$), charge transfer at the patterned anode ($P_{A\beta}$) and the cermet electrode (counter electrode) ($P_{A\gamma}$), and b) impedance spectrum and corresponding CNLS fit recorded at T = 800 °C, $pH_2 = 1.0\cdot10^3$ Pa, $pH_2O = 5.8\cdot10^3$ Pa, balance N_2, β_{rel} = 85 %. The individual processes $P_{A\alpha}$, $P_{A\beta}$, and $P_{A\gamma}$ are indicated by the dotted line. The spectrum is corrected for the inductance of the setup of $L = 4\cdot10^{-7}$ H.

table 3.3: List of the processes taking place in patterned Ni anodes, together with the values for the real part of the respective resistance, the relaxation frequency determined by $f_{rel} = (2\pi\cdot\tau)^{-1}$ and the exponent n according to $Z_{RQ} = r/(1+(j\omega)^n\cdot RQ)$. Additionally, the temperature and frequency dependencies and the attributed physical origin are given. Note that for the charge transfer process $P_{A\beta}$, both values of ASR and LSR are indicated.

Abbr.	**f_{rel} [Hz]**	**ASR [Ω·cm²]**	**LSR [Ω·m]**	**n**	**Dependency**	**Electrode process / physical origin**
$P_{A\alpha}$	1 to 6	1 to 20		0.95	pH_2O, pH_2	Gas diffusion
$P_{A\beta}$	0.1 to 100	5.5 to $3.4\cdot10^3$	36 to $3\cdot10^4$	0.9 to 0.95	pH_2O, pH_2, T	Charge transfer reaction at patterned Ni anode
$P_{A\gamma}$	$5\cdot10^3$ to 10^4	< 1		0.7 to 0.78	pH_2O, pH_2, T	Charge transfer reaction in Ni/8YSZ cermet electrode

Additionally, ohmic and inductive contributions have to be considered in the equivalent circuit model. The total ohmic contribution is attributed to the electrolyte (R_{ohm}) and the measurement setup $R_{setup}(800°C) = 1.3\ \Omega$. R_{setup} was determined by a short circuit measurement and shows a negligible temperature dependency. L accounts for the inductance of the setup (i.e. the wiring) with $L = 4{\cdot}10^{-7}$ H.

The separation of the processes $P_{A\alpha}$ and $P_{A\beta}$, i.e. gas diffusion and charge transfer, is favorably performed at T = 800 °C. The gas diffusion process is almost temperature independent and the DRT calculation yields very stable numerical results for the charge transfer process in this temperature regime. This separation is important as it allows for the first time to determine the LSR_{ct} instead of the overall LSR of patterned anodes.

3.4 Impurity Analysis

In this work, a detailed impurity analysis has been performed on patterned Ni anodes after the electrochemical characterization. This characterization included various parameter conditions of temperature, partial pressures of H_2 and H_2O and polarization voltage. The combination of SEM with several surface analysis tools, including Scanning Electron Microscopy (SEM), Atomic Force Microscopy (AFM), X-ray Photoelectron Spectrometry (XPS) and Time-Of-Flight Secondary Ion Mass Spectroscopy (TOF-SIMS) allows to locate and quantify impurity features and to determine their size.

Two different types of samples have been analyzed: (i) patterned Ni anodes after the electrochemical characterization and (ii) patterned Ni anodes after the characterization on which the Ni electrode has been removed by an acidic solution of HNO_3. Hence it is possible to assess all regions of interest: (i) the YSZ surface, (ii) the Ni surface, (iii) the electrode-electrolyte interface between YSZ and Ni, and (iv) the Triple-Phase-Boundary (TPB).

SEM

The SEM analysis was performed using a Zeiss 1540 XB (Carl Zeiss GmbH, Germany).

SEM images of a patterned Ni anode after characterization in varying H_2-H_2O atmospheres at 800 °C for 200 to 400 h are shown in figure 3.10: at high resolution, traces of impurities at the TPB are observable. In related literature on impurities at the TPB of point anodes, the notation of 'impurity ridges' is introduced for features with a size of up to a few microns in height [22]. With respect to the small dimensions of the features found in the present study, the notations 'impurity striation' and 'impurity imprint' seem more adequate.

The image of the sample with Ni electrode (figure 3.10a) shows impurity striations at the TPB. The width of these impurity striations differs between 100 nm to 1.6 μm, composed of adjacent lines starting from an almost straight line to a curved line next to the TPB. The individual lines are clearly distinguishable. With increasing distance from the TPB, the lines become more pronounced.

The images of the sample where the Ni electrode has been removed after electrochemical characterization are shown in figure 3.10b and c. They reveal impurity imprints of the Ni grain boundaries on the former electrode-electrolyte interface. No impurities are observable at the YSZ grain boundaries. Furthermore, no change in the electrolyte surface appearance, e.g. the hill-and-valley structure as reported in Ni point anode literature [22], is observable.

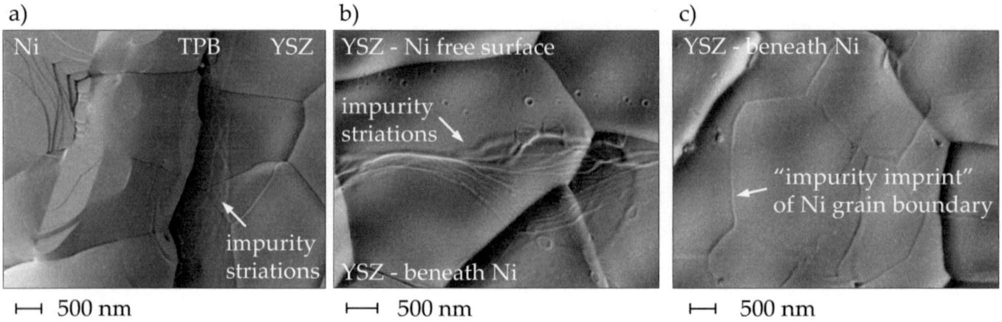

figure 3.10: SEM images of the patterned Ni electrode sample with Ni electrode after electrochemical characterization at 800 °C and varying H_2-H_2O atmospheres: a) with Ni electrode showing impurity striations close to the TPB, and b) and c) after removal of the Ni electrode with acidic solution. The varying width of the TPB impurity striation and the impurity imprints on the electrode-electrolyte-interface at the former Ni grain boundaries are visible.

AFM

For the AFM analysis, a DualScope atomic force microscope (DME Danish Micro Engineering A/S, Denmark) was used in tapping mode. The image resolution was 256 x 256; in combination with the appropriate tip this yields a height resolution of 9 pm.

The AFM analysis has been performed on the sample without the Ni electrode. High resolution and small area scans are necessary to detect the small impurity striations at the TPB and the former electrode-electrolyte interface, see figure 3.11. The broad impurity feature at the TPB is composed of several smaller adjacent striations. The height is determined to values in the range of 8 to 15 nm. An individual striation is about 10 to 30 nm wide, whereas the whole feature can extend up to 1.6 μm. The impurity imprints at the electrode-electrolyte interface are even smaller, they have an average height of 3 to 7 nm and are about 10 nm wide.

The detection of the impurities is extremely difficult as the curvature of the electrolyte surface is typically 150 to 250 nm and therefore exceeds the elevation of the striations by far. Furthermore, their small size makes determining the composition of the impurity feature almost impossible, as e.g. the XPS analysis covers a bigger area, with typical minimum spot sizes of 20 – 30 μm.

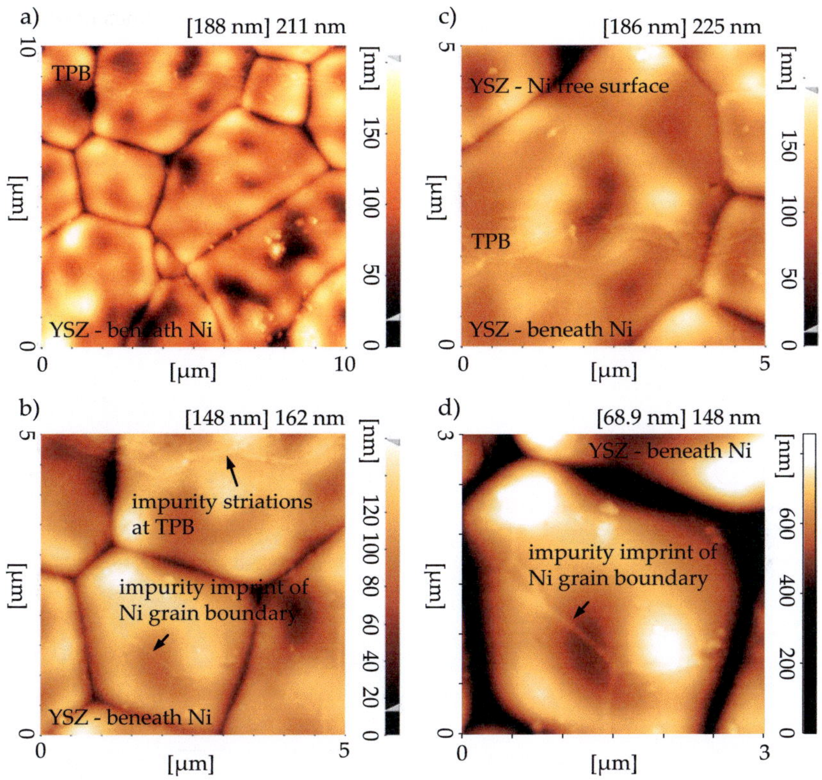

figure 3.11: Results of AFM analysis on the sample without Ni electrode (removed after electrochemical characterization at 800 °C in H_2-H_2O atmosphere). Very small impurity striations at the TPB and impurity imprints on the electrode-electrolyte interface at the former Ni grain boundaries are observable. The curvature of the YSZ surface of 150-250 nm exceeds the dimensions of the impurity features of 3-15 nm by far.

TOF-SIMS

TOF-SIMS was performed on a TOF-SIMS IV (ION-TOF GmbH, Germany) at a pressure of $4{\cdot}10^{-8}$ torr. High mass resolution spectra (low lateral resolution) were obtained using 25-ns pulses of 25-keV Bi^+ (primary ions), which were bunched to form ion packets with a nominal temporal extent of < 1 ns at a repetition rate of 10 kHz, thus yielding a target current of 0.9 pA. High lateral resolution images with about 250 nm (low mass resolution) were obtained using 160-ns pulses of 25-keV Bi^+ at a repetition rate of 10 kHz, thus yielding a target current of 150 fA. The data was post-processed using the TOF-SIMS IV software package (version 4.1). For a more detailed description of the procedure see [73].

The TOF-SIMS analyses have been performed on the surface of samples with and without Ni electrode. Two different areas with different lateral resolution have been chosen: (i) 150 x 150 μm^2 with high mass resolution and low lateral resolution, and (ii) 60 x 60 μm^2 with high lateral resolution and low mass resolution. These surface sizes ensure that several Ni stripes and YSZ surface stripes are covered by the scan. The detected species include the main components of the sample (Ni, Y and Zr) as well as the following impurity species: Na, Mg, Al, Si, K, Ca, Ti and Mn. A different lateral distribution corresponding to the Ni electrode stripe pattern could be observed for both sample types, see figure 3.12 and figure 3.13. By analyzing the intensity profiles of samples with Ni electrode and without Ni electrode, the location of the different impurities can be determined, see figure 3.14. Note that the signal intensity does not give quantitative information for different species. Furthermore, due to matrix effects different signal intensities may be detected for the same element on different substrates [73].

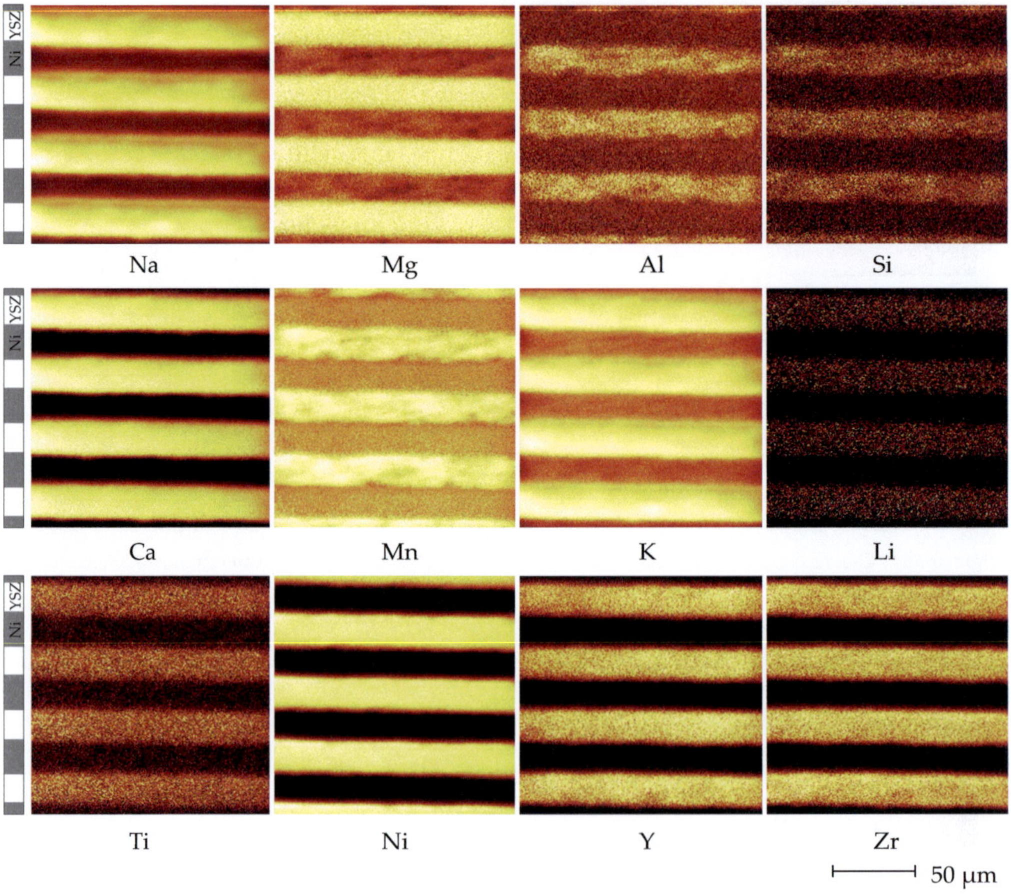

figure 3.12: TOF-SIMS ion images (150 x 150 μm²) of the surface of a sample with Ni electrode (stripe width 20 μm, spacing 20 μm) after electrochemical characterization at 800 °C in varying H_2-H_2O atmosphere. High mass resolution and low lateral resolution was used. Black is the lowest and white the highest ion intensity measured. The signal intensity of all detected species is different between Ni and YSZ surface, indicating the preferential location of the species.

On the sample with Ni electrode, Mn, Al and Si show a higher intensity on the Ni surface than on the YSZ surface, whereas Na, Mg, K, Ca, and Ti show a higher intensity on the YSZ electrolyte. However, this difference in signal intensity does not necessarily indicate a difference in concentration but may also be attributed to matrix effects. Interestingly, on the sample with Ni electrode, the higher Si intensity is found on Ni (see figure 3.14a), whereas on the sample without Ni electrode, the higher Si intensity is found on the YSZ surface (and not on the former interface). This indicates that the electrode-electrolyte interface is not favorable for Si segregation. Ti shows also a weaker intensity at the former electrode-electrolyte interface. Ca and Mg have a constant signal on the sample without Ni, indicating that the whole electrolyte is covered with these species even at the electrode-electrolyte interface, whereas Mn is mainly located at the electrode-electrolyte interface.

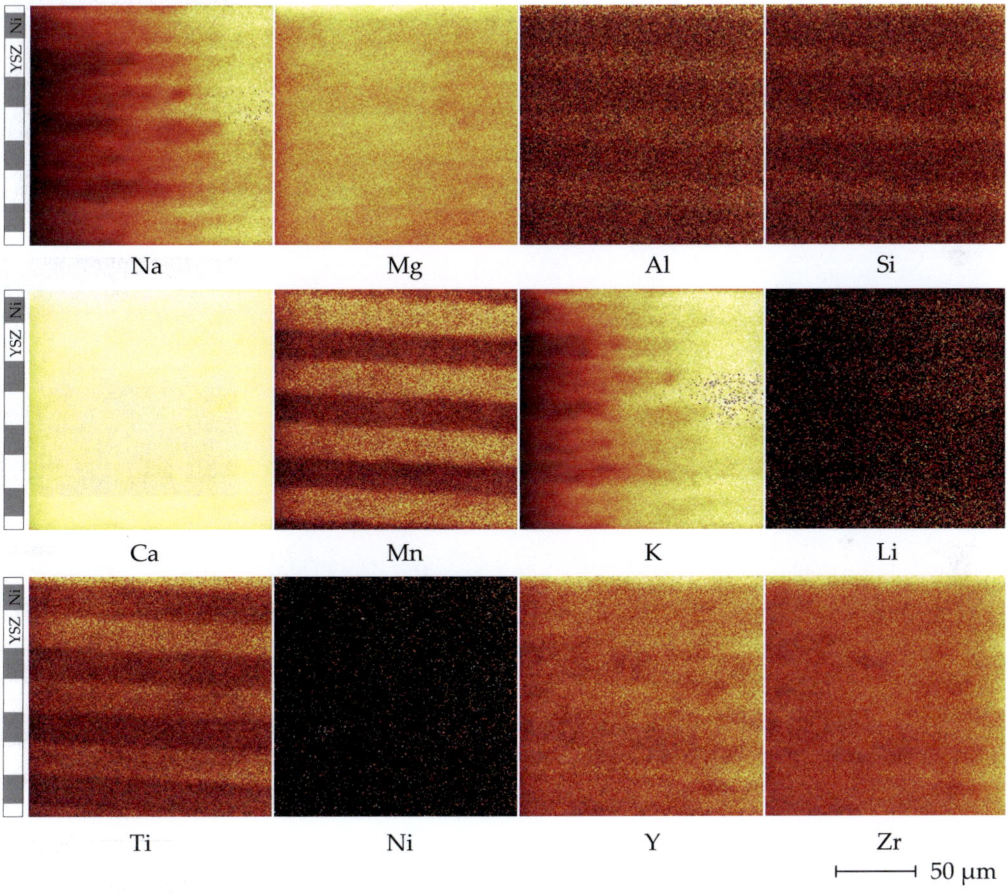

figure 3.13: TOF-SIMS ion images (150 x 150 μm²) of the surface of a sample where the Ni electrode (stripe width 20 μm, spacing 20 μm) has been removed after electrochemical characterization at 800 °C in varying H_2-H_2O atmosphere. High mass resolution and low lateral resolution was used. Black is the lowest and white the highest ion intensity measured. Only some species (Al, Si, Mn, Ti) show the former stripe pattern, whereas the others are equally distributed on the entire surface (Na, Mg, Ca, K). As expected, no Ni signal is detected.

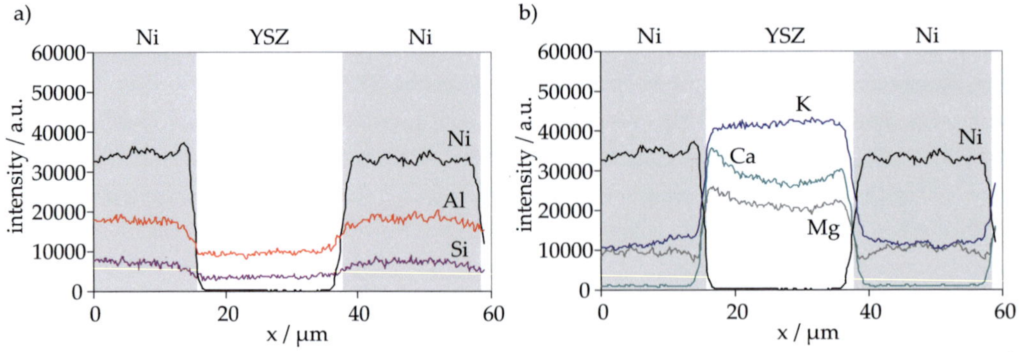

figure 3.14: TOF-SIMS line intensity profile of individual species compared to Ni signal taken from the 60 x 60 µm² TOF-SIMS scan of the sample with Ni electrode: a) impurity profiles of Al and Si. Both species show a higher intensity on the Ni surface than on the YSZ surface, and b) intensity profiles of K, Ca and Mg. Ca and Mg show an increase in intensity close to the Ni electrode and might indicate species accumulation at the impurity striations. However, the lateral extension of ca. 4 µm is rather large compared to an impurity feature width of 1 µm determined from AFM.

In figure 3.14b the line intensity profile of a 60 x 60 μm^2 scan of the sample with Ni electrode is shown. Whereas K shows constant signal intensity on the width of the YSZ surface, the signal intensity of Ca and Mg increases towards the TPB and may be an indicator for accumulation of these species at the TPB. However, the peaks have a width of about 4 µm, whereas the AFM analysis suggests a width of 1 µm for the TPB impurity striations. Another possible explanation for the increase in signal intensity is a shading effect of the Ni electrode on the signal intensity.

XPS

Chemical surface analysis was performed on a K-Alpha XPS (Thermo Fisher Scientific, UK), using a spot of monochromatized AlKα X-rays and employing charge compensation during measurements. Survey spectra were recorded using a pass energy of 150 eV. For the analysis, a spot size with a width of 200 µm width was used (400 µm for the sample without electrode) and a total of 25 scans have been chosen. Data analysis was performed using the software Thermo Avantage (Version 4.53).

The XPS analyses have been performed on the surface of a series of YSZ electrolytes: (i) as shipped, (ii) after sintering of the Ni/8YSZ cermet electrode at T = 1300 °C, and (iii) after characterization at 800 °C in H_2-H_2O atmosphere. A series of Ni electrodes has been analyzed as well: (i) after fabrication of the patterned Ni anode on the sintered YSZ substrate (before the electrochemical characterization), and (ii) after electrochemical characterization at 800 °C in H_2-H_2O atmosphere.

Besides the sample components (Ni, Y, Zr) and unavoidable C (due to exposure to air), the following impurity species were detected: Mg, Na, Si, Al, Mn, Ti, Zn and Fe. Ca could not be detected in the spectra as potentially present Ca peaks come very close to the peaks of Zr.

In table 3.4, the concentrations of all detected species are listed, and table 3.5 lists the ratio of selected species compared to the Zr signal. Additionally, the results are compared to values reported by Norrman et al. for the YSZ surface of 'pure' and 'impure' Ni point electrodes (last two columns) [73].

table 3.4: Results of XPS analysis with amount of main components (Y, Zr, O, Ni) and impurity species in at. %. The sum of all elements is 100 %.

Species	YSZ as shipped	YSZ after sintering (T=1300 °C)	YSZ after char. 800 °C, H_2-H_2O	YSZ after char. electrode removed	Ni before char.	Ni after char. 800 °C, H_2-H_2O
Zr	23.33	17.37	23.11	18.1	13.31	15.57
Y	4.80	3.26	5.03	3.97	2.46	2.78
O	57.89	48.44	58.28	47.03	44.57	49.79
Ni	-	-	-	-	9.04	8.10
C	7.46	26.68	7.76	24.71	18.50	12.69
Mg	2.11	0.27	1.59	4.63	0.11	0.39
Na	0.02	0.76	0.46	0.33	0.43	0.38
Si	4.04	2.05	0.57	0.49	1.27	1.78
Al	0.35	1.18	0.07	-	0.71	0.87
Mn	-	-	2.78	0.15	6.50	4.66
Ti	-	-	0.35	0.17	-	-
Zn	-	-	-	0.42	-	-
Fe	-	-	-	-	3.11	3.00

table 3.5: Ratio of selected elements on the YSZ substrate compared to the Zr signal. Additionally, the results are compared to values reported by Norrman et al. for 'pure' (99.995 %) and 'impure' (99.8 %) Ni point electrodes at the free YSZ surface [73]. 'Σ imp/Zr' refers to the sum of all impurity species excluding C signal.

Species ratio	YSZ as shipped (T=1300 °C)	YSZ after sintering (T=1300 °C)	YSZ after char. 800 °C, H_2-H_2O	YSZ after char. electr. removed	YSZ 'pure' Ni electr. [73]	YSZ 'impure' Ni electr. [73]
Y/Zr	0.21	0.19	0.22	0.22	0.37	0.35
Mg/Zr	0.090	0.016	0.069	0.256	-	-
Na/Zr	0.001	0.044	0.020	0.018	0.19	0.26
Si/Zr	0.173	0.118	0.025	0.027	0.35	0.6
Al/Zr	0.015	0.068	0.003	-	-	-
Mn/Zr	-	-	0.12	0.01	-	-
Ti/Zr	-	-	0.015	0.009	-	-
Σ imp/Zr	0.28	0.25	0.25	0.34	0.54	0.86

No increase in impurity content is observed during sintering of the electrolyte, which is an indication of a very pure electrolyte material and no contamination during sintering. Furthermore, the amount of impurities is stable during electrochemical characterization at 800 °C. The amount of Si and Al even decreases, whereas the species Mn and Ti appear during characterization, possibly segregating from the Ni electrode. The overall impurity amount with a ratio compared to the Zr signal of 0.28 is very small and significantly smaller than the result of 0.54 reported for point anodes with pure Ni electrode (99.995 % Ni) [73], see table 3.5. The same is valid for the Si/Zr signal with 0.025 after characterization compared to 0.35 reported by Norrman. This is of particular importance as SiO_2 is considered blocking the electrolyte surface.

The Y/Zr signal for the YSZ electrolyte with values of 0.19 to 0.22 is slightly higher than the signal ratio of 0.18 obtained for a polished electrolyte surface, which is assumed as composition of the YSZ bulk. This indicates that a minor demixing from the original composition of 8 mol % Y_2O_3 to 9 mol% Y_2O_3 has occurred near the surface.

The positioning of the XPS spot on the Ni electrode proved to be difficult. Therefore, the spectra contain also signals of Y and Zr. The Ni electrode shows a slightly higher impurity content than the YSZ electrolyte. This is expected as the purity of the Ni target is only 99.98 %. The major impurity component is Mn, which is not listed in the chemical analysis of the Ni sputter target delivered by the supplier, see section 3.1. During the electrochemical characterization, the Si amount increases slightly, but remains very small with 1.27 and 1.78 a.% before and after characterization respectively.

Discussion

The surface analysis of the patterned Ni anodes revealed small impurity features at the TPB and on the electrode-electrolyte interface at the former Ni grain boundaries. Furthermore, the TOF-SIMS analysis showed different surface compositions on the free YSZ surface, the Ni surface and the electrode-electrolyte interface. The size of the impurity striations at the TPB with a height of up to 20 nm is comparable to values reported for very pure point anodes, whereas heights exceeding 1 μm were observed for less pure Ni electrodes [23]. Herein, a height of 20 nm corresponds to about 40 atomic layers of SiO_2. In the case of a direct location at the TPB, this would mean a barrier for the ion transfer reaction and favor a reaction mechanism involving interstitial transport of hydrogen, as proposed in earlier studies with point anodes [26, 63, 71]. However, for the patterned anodes in this work it is assumed that, due to the initial shrinkage of the Ni pattern, this impurity striation is not exactly located at the TPB, as will be discussed later. In any case, this impurity level can be considered as the best achievable, that was enabled by the careful choice of materials and a measurement setup that avoids further contamination. Especially the amount of the critical impurity element Si is very small, demonstrating that no contamination source was present in the sintering oven or the measurement setup.

A major difference between the present patterned electrodes and point electrodes by Høgh [67] is that SiO_2 and Al_2O_3 contamination was found on top of the Ni patterns and even with a concentration higher than on the YSZ. SiO_2 and Al_2O_3 were not detected on the Ni-wires by Høgh. The cause of this difference is not yet clear but may be due to impurities in the contact mesh (purity ≥ 99.6 %).

The lateral distribution of impurities at the electrode-electrolyte interface differs between the patterned Ni anodes of this work and results reported for point anodes: no distinction between a purer rim zone compared to an impure core zone as reported by Schmidt et al. [26, 71] was found. It is assumed that this is due to the differences between the two electrode types with respect to anode preparation and morphological development: While patterned anodes are joined permanently to the electrolyte, point anodes are realized by temporarily pressing a bent Ni wire on the electrolyte; proper contacting is ensured by a load. During electrochemical characterization, the Ni film of the patterned anode exhibits shrinkage of the stripe width caused by Ni grain growth, whereas the electrode-electrolyte interface of the point anode area increases due to Ni creep caused by the load. Assuming that segregation of impurities to the free YSZ surface is more favorable than to the electrode-electrolyte interface, the interface of patterned anodes should be less contaminated than the core zone of a point electrode. Based on the same assumption, the YSZ surface directly at the TPB is supposed to exhibit fewer impurities than the free YSZ surface, as it has originally been covered by Ni. This is in accordance with findings by Schmidt with a flattened electrode [71]: compared to a standard point electrode that showed an initial decrease in LSR during thermal exposure, which is ascribed to slow creep of electrode and formation of less impure rim zone, the flattened electrode showed an almost constant LSR from the beginning.

The assumption of a less contaminated Ni-YSZ interface could also explain the initial behavior observed for the patterned Ni anodes analyzed in this work and reported in [58]: the initial decrease in LSR is caused by an increase in l_{TPB} as well as a widening of the active reaction zone due to the shrinkage of the Ni electrode. The observed degradation could then be attributed to a slow, gas composition dependent diffusion of impurities closer to the TPB. This assumption is based on the increase in the production rate of volatile $Si(OH)_4$ with increasing pH_2O [86], which leads to a higher mobility of the impurities.

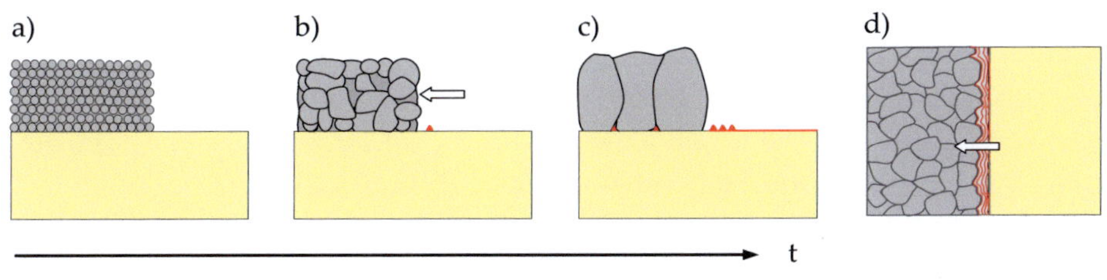

figure 3.15: Scheme for the development of the impurity film topography found at the TPB of patterned Ni anodes over time: a) after sputtering, the Ni film is composed of small grains, b) during thermal exposure the Ni grains grow in size. At the same time, first impurities segregate to the TPB originating either from Ni, YSZ or both, c) the Ni grains grow further and lead to a shrinkage of the Ni stripe width, yielding a broadening of the impurity film. It is assumed that the electrolyte surface in direct neighborhood of the electrode is less contaminated, d) the top view illustrates that the impurity film varies in width depending on the grain structure.

Based on these observations, the following model for the initial behavior of patterned Ni electrodes has been developed and is illustrated in figure 3.15. After sputtering, the Ni pattern is composed of very small grains (few nm); no segregation of impurities at the TPB has occurred at room temperature (figure 3.12a). During thermal exposure at 800 °C, impurities from electrode and electrolyte material segregate to the TPB and the free Ni and YSZ surfaces. At the same time, the Ni stripe width reduces due to grain growth, leaving a less contaminated surface directly at the TPB. The assumption is that the electrolyte surface directly adjacent to the TPB is not covered by an impurity striation and is even less contaminated than the free YSZ surface (figure 3.12b). With further shrinkage of the Ni stripe, several adjacent striations form next to the TPB. The observation that the first striation at the outermost position from the TPB is most pronounced indicates that segregation is very fast and occurs before the initiation of grain growth, e.g. already during heating (figure 3.12c and d). Another favorable place for segregation is at the Ni grain boundaries rather than at the YSZ grain boundaries.

Additionally, a theoretical calculation has been performed to determine the probability of the electrolyte surface being completely covered by bulk impurities, see section 9.2 and [87]. It has been found that, due to the unfavorable surface to volume ratio of patterned and point anodes, impurities at the surface are almost unavoidable for these anode concepts. Hence, it is important to consider size and possible effects of impurity features in the interpretation of the electrochemical performance.

3.5 Summary

In this chapter, the following characteristics of the patterned Ni anodes used in this work have been determined:

A reproducible fabrication method for patterned Ni anodes with sharp edges was introduced consisting of Ni sputtering on structured photoresist followed by lift-off of the photoresist.

The optimal Ni layer thickness was determined that ensures stability up to temperatures of 800 °C in reducing atmosphere. The initial morphological changes of the Ni film during thermal exposure stabilize after about 20 h at 800 °C. The correction factor τ_{corr} = 1.3 accounts for the corresponding increase in TPB length for H_2-H_2O and CO-CO_2 atmosphere. A standard initial heat treatment ensures comparability between different samples.

Typical errors induced by reference electrodes were avoided by the characterization against a Ni/8YSZ cermet counter electrode with known contribution.

The electrochemical characterization was performed in a gas tight measurement setup that allows a precise adjustment and monitoring of the gas composition. A large variation of water vapor contents of 0.02 % $\leq \beta_{rel} \leq$ 88 % was realized by an upstream combustion chamber.

The combined analysis of impedance data by DRT calculation and CNLS fits allows the separation of the charge transfer reaction at the patterned anode ($R_{A\beta}$) from contributions of gas diffusion ($R_{A\alpha}$) and counter electrode ($R_{A\gamma}$). Hence, for the first time the pure charge transfer contribution LSR_{ct} of patterned Ni anodes and its parameter dependencies have been determined.

Even though impurities cannot be avoided entirely for Ni electrodes at an elevated temperature of 800 °C, their amount has been kept at a minimum by the careful choice of materials and a measurement setup that avoids further contamination.

4 Investigation of Relaxation and Degradation Mechanisms

In this chapter, the reproducibility of the electrochemical characterization of patterned Ni anodes is addressed. So far, the strong deviations in existing literature on patterned anodes could only vaguely be attributed to impurities. However, no information on the impurity features for the different studies is available. Based on the suspicion that the reported experimental data is also subject to stability problems and degradation, a detailed analysis of the long-term behavior of patterned anodes has been performed.

By a systematic observation of the behavior of patterned Ni anodes, four different peculiarities could be described: (i) the initial decrease in LSR during thermal exposure is described in section 4.1, (ii) the degradation rate at constant conditions is described in section 4.2, (iii) the slow relaxation process upon changes in the gas composition is addressed in section 4.3, and (iv) the relaxation upon polarization is described in section 4.4.

4.1 Initial Behavior

In this work, only samples with a layer thickness of 800 nm have been characterized. This thickness, combined with a minimum stripe width of 5 µm, has been shown to ensure the stability of the Ni structure during electrochemical characterization at temperatures up to 800 °C in reducing atmosphere [51].

Nevertheless, the geometrically stable patterned Ni anode structure is still undergoing morphological changes during characterization. At 800 °C, the LSR_{ct} initially shows a considerable activation within the first 20 - 25 hours of thermal exposure, see figure 4.1. In this time range, a decrease in LSR_{ct} of 60 - 65 % is observed with subsequent stabilization at an average value of LSR_{ct} of 54 Ω·m (note that for this calculation $l_{TPB,th}$ has been chosen).

This reproducible initial activation of all samples is attributed to morphological changes in the Ni thin film (grain growth, agglomeration, surface roughness) that result in both an increase of l_{TPB} and a change of in-plane conductivity. After 20 - 25 h the morphological changes are almost entirely completed, and further alterations in l_{TPB} during characterization are negligible. This assumption is affirmed by the SEM images of a patterned Ni anode at different stages of the characterization, see figure 4.2. The grain growth and the consequential increase in l_{TPB} after the initial heat treatment for 20 h ($pH_2 = 8.6 \cdot 10^4$ Pa, $pH_2O = 1.5 \cdot 10^4$ Pa) is obvious, see figure 4.2a and b. For the calculation of LSR, this increase in l_{TPB} is accounted for by the correction factor τ_{corr}, which is the quotient of actual $l_{TPB,corr}$ and theoretical $l_{TPB,th}$. A value of $\tau_{corr} = 1.3$ was estimated using the image analysis tool SPIP [79]. Comparing figure 4.2b and c, no further increase in average grain size and l_{TPB} is observed and the same correction factor is obtained.

With respect to a possible effect of impurities, the strong decrease in LSR_{ct} could additionally be attributed to the creation of a 'clean' TPB area as a consequence of the shrinkage of the Ni electrode, see discussion in section 3.4.

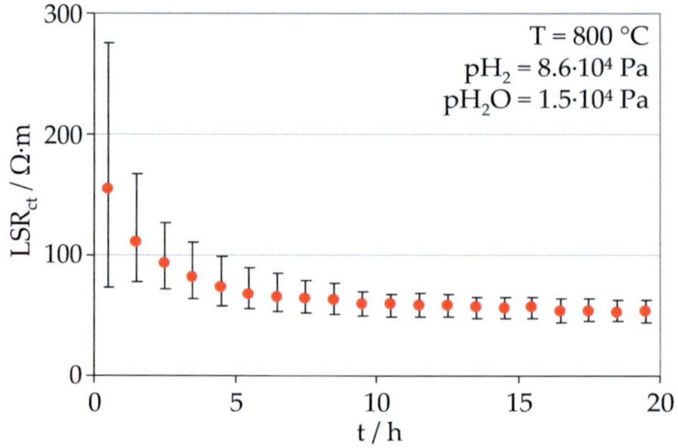

figure 4.1: Average LSR_{ct} value of all patterned Ni anodes with different l_{TPB} plotted over time during the first 20 h of thermal exposure at constant conditions, together with the error bars for minimum and maximum values obtained for the different samples (layer thickness: 800 nm, T = 800 °C, $pH_2 = 8.6 \cdot 10^4$ Pa, $pH_2O = 1.5 \cdot 10^4$ Pa, $\beta_{rel} = 15$ %).

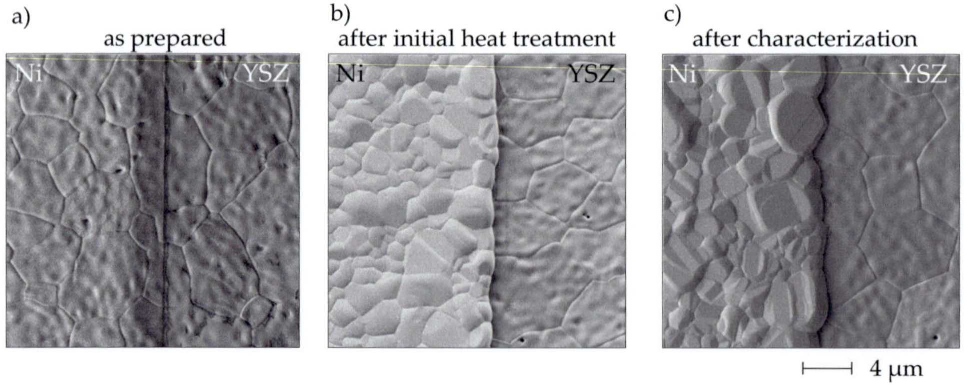

figure 4.2: SEM images of the patterned Ni anode at different stages of the electrochemical characterization: a) as prepared, b) after initial heat treatment at T = 800 °C, $pH_2 = 8.6 \cdot 10^4$ Pa, $pH_2O = 1.5 \cdot 10^4$ Pa, $\beta_{rel} = 15$ % for 20 h, and c) after electrochemical characterization at 700 °C $\leq T \leq$ 800 °C, $6.9 \cdot 10^3$ Pa $\leq pH_2O \leq 8.3 \cdot 10^4$ Pa, $1.8 \cdot 10^4$ Pa $\leq pH_2 \leq 8.7 \cdot 10^4$ Pa, 13 % $\leq \beta_{rel} \leq$ 82 %, balance N_2, exposure time = 210 h.

With respect to this initial behavior of patterned Ni anodes and the need of reproducible measurement data, the necessity of a standard initial heat treatment is apparent. It has been shown that the morphological changes of the Ni layer are strongly dependent on temperature, see section 3.1.3 and [51]. Therefore, it is important to perform the initial heat treatment at the highest temperature of interest for the electrochemical characterization in order to avoid a further increase in l_{TPB} after this initial stabilization. Throughout this work this was realized by the introduction of a standard initial heat treatment: before the actual measurement program, all samples were exposed to 800 °C and 15 % relative water vapor content ($pH_2 = 8.6 \cdot 10^4$ Pa, $pH_2O = 1.5 \cdot 10^4$ Pa, $pN_2 = 0$ Pa) for 20 h. To ensure a reproducible starting point for the characterization in CO-CO_2, the same standard initial heat treatment was applied.

4.2 Degradation Rate

After the first period of activation, a constant degradation of LSR_{ct} is attained as a function of gas composition. In H_2-H_2O atmosphere, a degradation rate of 0.4 % / h is observed at 800 °C and a high water vapor content ($pH_2 = 3.9 \cdot 10^4$ Pa, $pH_2O = 6.2 \cdot 10^4$ Pa, $pN_2 = 0$ Pa), see also figure 4.8. This is a very high degradation rate, exceeding the degradation rate of highly efficient Ni/8YSZ cermet anodes (0.004 % / h) [2] by two orders of magnitude. A comparison among patterned Ni anodes in literature is not possible, as none of the previous studies reported a degradation rate [6-9]. In CO-CO_2 atmosphere, a slower degradation rate of 0.18 % / h is observed at 800 °C and $pCO = 4.8 \cdot 10^4$ Pa, $pCO_2 = 5.4 \cdot 10^4$ Pa, see figure 4.3.

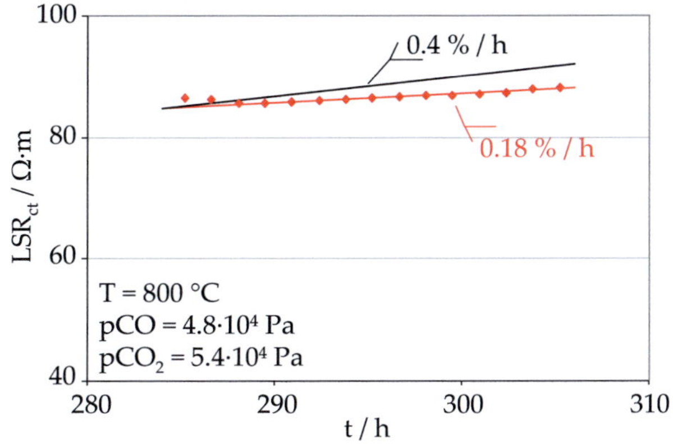

figure 4.3: Degradation behavior of a patterned Ni anode recorded at 800 °C for $pCO = 4.8 \cdot 10^4$ Pa and $pCO_2 = 5.4 \cdot 10^4$ Pa (sample: NiYSZ_291), together with the degradation rate obtained for characterization in H_2-H_2O atmosphere from figure 4.8.

The high degradation rate in H_2-H_2O atmosphere may be attributed to a removal of Ni via the gas phase e.g. by $Ni(OH)_2$. However, in CO-CO_2 atmosphere the patterned Ni anode is still subject to degradation, indicating an additional degradation mechanism. One assumption involves the diffusion of impurities to the TPB. In the study on impurity features in patterned Ni anodes, it has been suggested that the YSZ surface directly at the TPB is rather pure, due to the morphological changes of the Ni pattern during thermal exposure (i.e shrinkage), see section 3.4

and [87]. As soon as the stripe width stabilizes, no new pure YSZ surface at the TPB is uncovered by the shrinking stripe. Instead, impurities will start segregating to the TPB e.g. by surface diffusion. As the properties of a glassy silicate are strongly dependent on the water vapor content, this may additionally explain the gas composition dependent degradation rate observed for H_2-H_2O atmosphere.

4.3 Relaxation upon Changes in Gas Composition

The systematic observation of the degradation behavior additionally enabled the assessment of relaxation processes with large time constants taking place during the characterization of patterned Ni anodes. Consecutive impedance measurements in intervals of 60 min recorded after each gas composition change reveal reversible relaxation processes of LSR with long settling times for the H_2-H_2O atmosphere, see figure 4.4. The behaviors upon a variation of pH_2 and pH_2O vary significantly: whereas settling times for changes in pH_2O are 2 - 3 h, settling times following pH_2 variations exceed 5 h for most gas compositions at 800 °C, and even increase at lower temperatures. For pH_2O, the direction of relaxation depends on the direction of gas change, for pH_2 the correlation is more complex. On the one hand, the direction of relaxation depends on the actual pH_2: above a certain pH_2 (here $3 \cdot 10^4$ Pa) LSR decreases, below this value LSR increases. The amount and the settling time of the relaxation process seem to be inversely proportional to the actual pH_2 value. On the other hand, the recording of rapid series of pH_2 variation with only 5 min delay at each gas composition indicates that the actual LSR value additionally depends on the direction of gas change: at a certain pH_2, two different LSR values are measured depending on the previous pH_2 (higher value for increasing pH_2), see figure 4.5b. Such a hysteresis is not observed in the case of rapid pH_2O variation, see figure 4.5a.

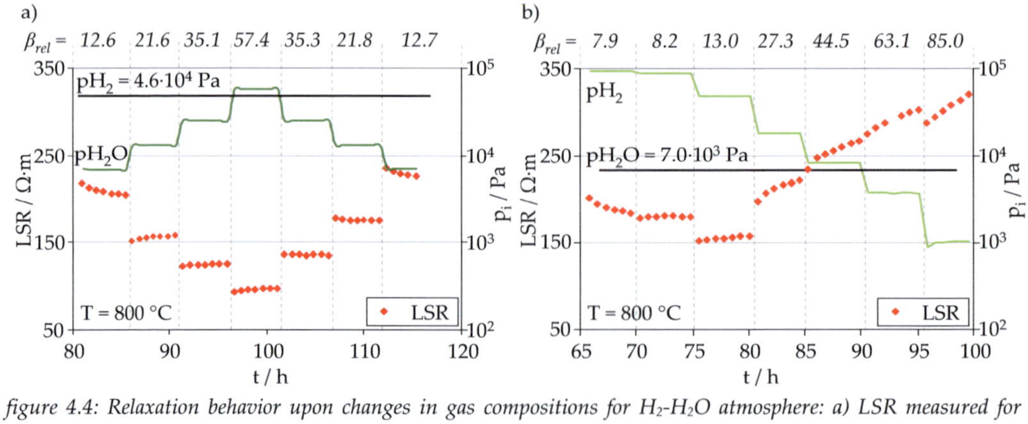

figure 4.4: Relaxation behavior upon changes in gas compositions for H_2-H_2O atmosphere: a) LSR measured for pH_2O variations at constant $pH_2 = 4.6 \cdot 10^4$ Pa (sample: NiYSZ_143), and b) LSR measured for pH_2 variation at constant $pH_2O = 7.0 \cdot 10^3$ Pa (sample: NiYSZ_146). β_{rel} indicates the relative water vapor content of the respective gas composition in percent. (T = 800 °C, balance: N_2).

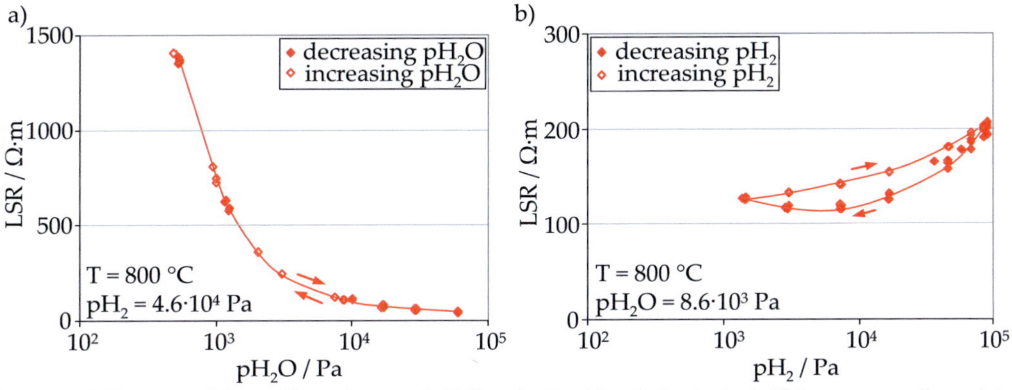

figure 4.5: LSR measured for rapid gas changes of pH_2O and pH_2 with only 5 minutes hold time per operation point: a) no hysteresis observable for pH_2O variations ($pH_2 = 4.6\cdot10^4$ Pa), and b) considerable hysteresis observable between increase and decrease of pH_2 ($pH_2O = 8.6\cdot10^3$ Pa). T = 800 °C, balance: N_2, sample: NiYSZ_148.

The corresponding measurement with 5 hours delay time per gas composition for CO-CO_2 atmosphere is shown in figure 4.6. No elongated settling times as for pH_2 and pH_2O changes are observed, instead equilibrium is typically attained within one hour. Furthermore, no differences were found between the behavior for pCO and pCO_2.

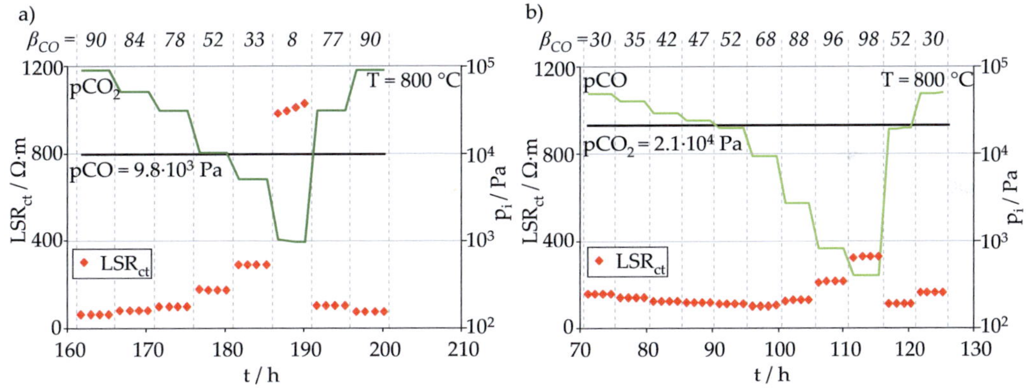

figure 4.6: Relaxation behavior upon changes in gas composition for CO-CO_2 atmosphere: a) LSR measured for pCO_2 variations at constant $pCO = 9.8\cdot10^3$ Pa, and b) LSR measured for pCO variation at constant $pCO_2 = 2.1\cdot10^4$ Pa. For each gas composition, $\beta_{rel,CO/CO2}$ is given in percent. (T = 800 °C, balance: N_2, sample: NiYSZ_259).

In figure 4.7, LSR values for pH_2 and pCO variation are compared at the same scale to illustrate the strong difference in relaxation behavior. From these observations it is evident that the parameter variation series for pH_2 variation are strongly affected by the delay between a change in gas composition and the actual recording of impedance spectra.

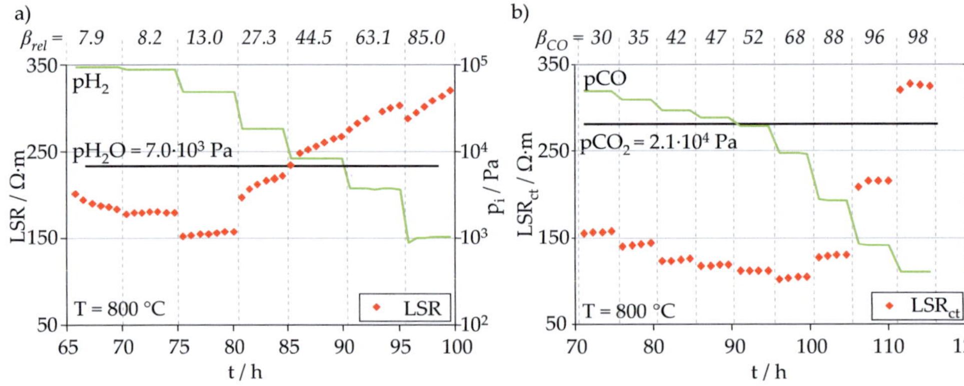

figure 4.7: Comparison of relaxation behavior upon gas composition changes for H_2-H_2O and CO-CO_2 atmosphere: a) LSR measured for pH_2 variation at constant $pH_2O = 7.0 \cdot 10^3$ Pa (sample: NiYSZ_146), and b) LSR measured for pCO_2 variation at constant $pCO = 9.8 \cdot 10^3$ Pa (sample: NiYSZ_259). For each gas composition β_{rel} and $\beta_{rel,CO/CO2}$ are given respectively (T = 800 °C, balance: N_2,).

Gas phase processes can be excluded as a cause for the long settling times, as a constant Nernst potential is attained within few minutes after a change in gas composition. With respect to the long settling times of 2 to 5 hours, surface processes such as adsorption, diffusion or chemical reactions seem improbable too, as they are expected to proceed much faster, see e.g. the high mobility of H on Ni [34]. A probable explanation involves a slow settling of species concentration in the bulk of the Ni electrode or the YSZ electrolyte, for example hydrogen atoms in Ni or interstitial protons in YSZ [88, 89]. Contrary to the high solubility of hydrogen in Ni, CO is not soluble in Ni and YSZ due to the large molecule size.

Thus, the long settling times in the case of H_2-H_2O atmosphere may be interpreted as a slow equilibration between bulk, surface and gas phase, whereas no equilibration takes place for CO-CO_2 and consequently settling times are fast. At the same time, this is an indication of a rate limitation including interstitial transport in the case of H_2-H_2O atmosphere.

4.4 Relaxation upon Recording of Current-Voltage Curves

After the recording of current-voltage curves, a reproducible drop in LSR of 20 % is observed, followed by a slow relaxation process with a settling time of about 20 h, see figure 4.8. This effect is reproducible and increases with repeated recording of current-voltage curves. The corresponding current-voltage curve for $-350 \text{ mV} \leq \eta \leq 400 \text{ mV}$ is shown in figure 4.9. A considerable hysteresis is observable between rising and falling polarization voltage for anodic and cathodic polarization. For both polarizations, the current density increases between rising and falling voltage, indicating that the resistance is activated by the applied overpotential. Hence, a reproducible short term activation of patterned anodes takes place during the recording of a current-voltage curve. This activation is followed by a slow relaxation process with settling times of about 20 h.

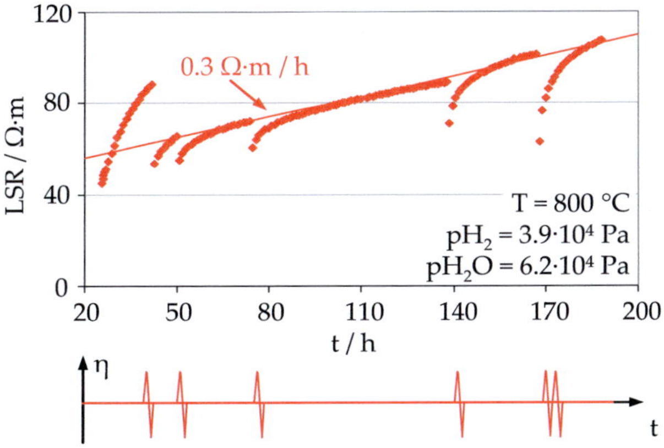

figure 4.8: LSR measured at constant conditions interrupted by current-voltage curves with -350 mV ≤ η ≤ 350 mV (see lower part). A slow relaxation after each recording of a current-voltage curve is observed as well as a linear degradation of 0.3 Ω·m / h = 0.4 % / h. ($pH_2 = 3.9\cdot10^4$ Pa, $pH_2O = 6.2\cdot10^4$ Pa, β_{rel} = 61.4 %, T = 800°C, sample: NiYSZ_151).

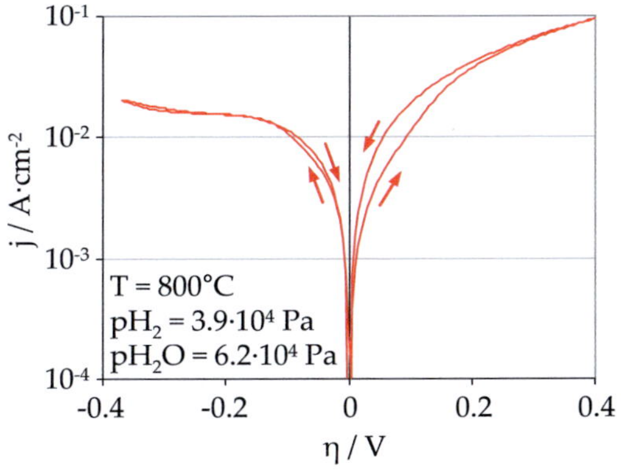

figure 4.9: Typical current-voltage curve recorded during the measurement with constant conditions from figure 4.8. Arrows indicate the directions of rising and falling voltages (speed of recording: 2 mV·s^{-1}). A considerable hysteresis is observed especially for anodic polarization.

Temperature activation as a consequence of current passage can be excluded as explanation for the hysteresis as the currents are too small. Again, the most likely explanation involves a temporary change in the (Ni or YSZ) bulk species concentration as already discussed for the relaxation processes after gas variation.

For point anodes, similar dynamics have been observed and additional oscillations of the current density at constant potential have been reported by Hansen at T = 1000 °C on the scale of few hours for Ni purity of 99.8 % [65] as well as by Schmidt at T = 850 °C on the scale of few minutes [72]. In this work, no such oscillations of the current were observed. However, the suggested explanations are of interest. Schmidt assumes a periodical change between (i) activation due to local increase in pH_2O, (ii) passivation of the Ni surface by $Ni(OH)_2$, slowing down the oxidation reaction, and (iii) reduction of $Ni(OH)_2$ as pH_2O decreases. Hansen explains the instabilities in current density by changes in the chemical composition of impurities and a physical redistribution of impurities [65, 68]. In a similar way, Høgh explains the observed decrease in polarization resistance after high cathodic polarization by a reduction of the impurity species at the TPB [67].

4.5 Conclusions

This chapter investigated the reproducibility of the electrochemical characterization of patterned Ni anodes. It was found that patterned Ni anodes are undergoing considerable relaxation and degradation processes with time constants in the order of several hours:

(i) An initial decrease of LSR_{ct} by 60 % to an average value of 54 $\Omega \cdot m$ during the first 20 h of thermal exposure at 800 °C was observed. This process is verified by SEM studies and attributed to morphological changes in the Ni thin film.

(ii) A degradation rate as a function of gas composition was observed. A value of 0.4 % / h was determined for a high water vapor content ($pH_2 = 3.9 \cdot 10^4$ Pa, $pH_2O = 6.2 \cdot 10^4$ Pa, $\beta_{rel} = 61$ %) at 800 °C. In CO-CO_2 atmosphere, a lower degradation rate of 0.18 % was observed. It is assumed that the degradation is caused by the formation of gaseous $Ni(OH)_2$ and eventually the diffusion of impurities to the TPB.

(iii) A slow relaxation process after variations of pH_2 and pH_2O was discovered. For pH_2O variations, settling times are 2 - 3 h at 800 °C, while pH_2 variations necessitate longer settling times of 5 - 8 h at 800 °C. For CO-CO_2 atmosphere, equilibrium values were attained within few minutes. Therefore, it is assumed that this behavior is coupled to variations in hydrogen concentration in the bulk of electrode and electrolyte.

(iv) A reproducible sequence of fast activation of cell performance during the recording of current-voltage curves followed by a slow relaxation process with settling times of about 20 h is observable.

In order to assure the reproducibility of characterization results, a standardized initial heat treatment was established, a sufficient delay (several hours) between parameter variation and measurement was applied and the degradation rate especially at high water vapor contents was considered. Hence, this is a significant improvement compared to previous studies on patterned Ni anodes, which did neither (i) report these slow processes, nor (ii) specify the entire measurement procedure.

5 Electrochemistry of the H_2 Oxidation

This chapter summarizes the results of the electrochemical characterization of patterned Ni anodes in H_2-H_2O atmosphere.

The Line Specific Resistance (LSR) and its parameter dependencies give evidence on the reaction mechanism of the electrochemical oxidation of H_2 [56]. Thus, a comprehensive data set of the dependencies on gas composition, temperature and polarization voltage represents the basis for the validation of elementary kinetic modeling studies. With respect to already existing experimental data, special focus needs to be directed to the identification of deviations and their explanation.

The experimental setup is explained in section 5.1. In section 5.2, the analysis approach with a description of the parameter dependencies according to the Butler-Volmer equation is given, together with the resulting parameter dependencies. A detailed comparison to literature data is performed in section 5.3, and to point anodes with known impurity features in section 5.4. Additionally, the influence of the double layer on the capacitive behavior of patterned Ni anodes is addressed in section 5.5.

5.1 Measurement Setup

For the electrochemical characterization described in this chapter, the sample design was composed of parallel stripes. Two perpendicular Ni stripes at the ends ensure proper contacting of all stripes. For details to the fabrication procedure and the contacting of the samples see chapter 1. For the measurement, the setup described in the same chapter in figure 3.4a has been chosen. The contacting is realized by a Ni mesh on the entire area of the Ni pattern.

Electrochemical impedance spectra (EIS) and current-voltage (C/V) curves were recorded over an extended range of gas compositions: $8 \cdot 10^2$ Pa $\leq pH_2 \leq 9 \cdot 10^4$ Pa and $2 \cdot 10^1$ Pa $\leq pH_2O \leq 6 \cdot 10^4$ Pa. As shown in chapter 1 in figure 3.6a, three series of gas variation of pH_2 and pH_2O respectively were realized with up to 7 points per series; N_2 was used as balance gas. This corresponds to a variation in relative water vapor content of 0.02 % $\leq \beta_{rel} \leq$ 88 %. Further parameter variations comprise temperature 450 °C $\leq T \leq$ 800 °C and polarization voltage -300 mV $\leq \eta \leq$ 300 mV.

The recording of impedance spectra was performed both by a Zahner IM6 and a Solartron 1260 FRA, in a frequency range of 100 mHz – 1 MHz. Due to the high impedance of the patterned anodes voltage stimulus was chosen instead of current stimulus; the amplitude was 10 mV.

5.2 Results of Cell Measurements

The following results were recorded considering the relaxation and degradation mechanisms described in chapter 4. For the description of the parameter dependencies of LSR_{ct} on gas composition, temperature and polarization voltage the Butler-Volmer equation is applied, see equation 5.1. In this way, comparability is enabled to previous studies with patterned Ni anodes [6-9] as well as to the study by Leonide et al. for Ni/8YSZ cermet anodes [90].

$$j = j_0 \left[\exp\left(\alpha \frac{n_e F \eta_{ct}}{RT} \right) - \exp\left(-(1-\alpha) \frac{n_e F \eta_{ct}}{RT} \right) \right] \tag{5.1}$$

Here j_0 is the (partial pressure and temperature dependent) exchange current density, n_e the number of exchanged electrons (in our case $n_e = 2$), α the charge transfer coefficient and η_{ct} the activation overpotential. The charge transfer coefficient α is an indicator of the symmetry of the activation energy barrier when a positive or negative overpotential is applied [91].

The exchange current density j_0 is described by a power law ansatz for the partial pressure dependencies of pH_2 and pH_2O and an Arrhenius-type temperature dependency [91, 92]:

$$j_0 = \gamma (pH_2)^a (pH_2O)^b \exp\left(-\frac{E_{act}}{RT} \right) \tag{5.2}$$

The Butler-Volmer equation does not consider the individual steps at the TPB as applied in modeling studies of the surface level, but represents a global approach that is typically used in cell level modeling [15]. It is based on the assumption of a single or few charge transfer steps with well defined energy barrier. Hence, it only applies if charge transfer is the rate determining step and no limitation or co-limitation of adsorption/desorption and surface transport persists [18, 91].

The current-voltage characteristics consist of a linear behavior of current to overpotential for small applied overpotentials and Tafel behavior for large overpotentials. The observed behavior of LSR_{ct} is reasonably well described by the Butler-Volmer equation and therefore this equation is applied in this work. The detailed modeling considering the individual elementary kinetic steps is subject of a separate study.

The parameter dependencies of LSR_{ct} are determined based on the inverse proportionality of exchange current density and resistance ($LSR_{ct} \sim 1 / j_0$) and are described by the exponents a and b (dependency of LSR_{ct} on pH_2 and pH_2O respectively), E_{act} (temperature dependency) and α (dependency on overvoltage), where c_1, c_2 and c_3 are proportionality constants:

$$LSR_{ct} = c_1 \left(pH_2\right)^{-a} \tag{5.3}$$

$$LSR_{ct} = c_2 \left(pH_2O\right)^{-b} \tag{5.4}$$

$$LSR_{ct} = c_3 \exp\left(\frac{E_{act}}{RT}\right) \tag{5.5}$$

In the following, the gas composition dependencies are presented in a double logarithmic plot with pH_2 or pH_2O as abscissa and LSR_{ct} as ordinate. With this presentation, the exponents a and b correspond to the negative slope, see equations 5.3 and 5.4. For the temperature dependency a plot of inverse temperature $1000 \cdot T^{-1}$ as abscissa and the logarithm of LSR_{ct} to base 10 as ordinate is chosen throughout this work, even though the activation energy E_{act} is calculated using the natural logarithm, see equation 5.5.

5.2.1 Partial Pressures of H_2O and H_2

The dependencies of LSR_{ct} on pH_2O and pH_2 are shown in figure 5.1 and figure 5.2 respectively. A strong dependency on pH_2O is observable with an exponent $b = 0.68$, whereas LSR_{ct} is almost independent of pH_2 variations with an exponent of $a = 0.07$. Additionally, in the characterization of pH_2 variation, a considerable scattering of the exponents between different samples was observed in the range of $-0.15 \leq a \leq 0.14$.

The qualitative pH_2O dependency determined in this work is in agreement with previous patterned Ni anode literature; good quantitative agreement is observed with data from Bieberle with $b = 0.67$ [6]. Data by de Boer and Mizusaki indicate a weaker dependency, see the literature review in chapter 2, table 2.2.

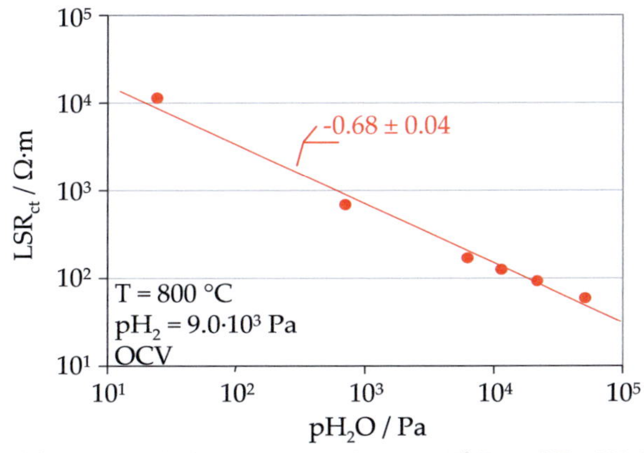

figure 5.1: LSR_{ct} measured for pH_2O variation at constant $pH_2 = 9.0 \cdot 10^3$ Pa and $T = 800$ °C. N_2 is used as balance, sample: NiYSZ_139.

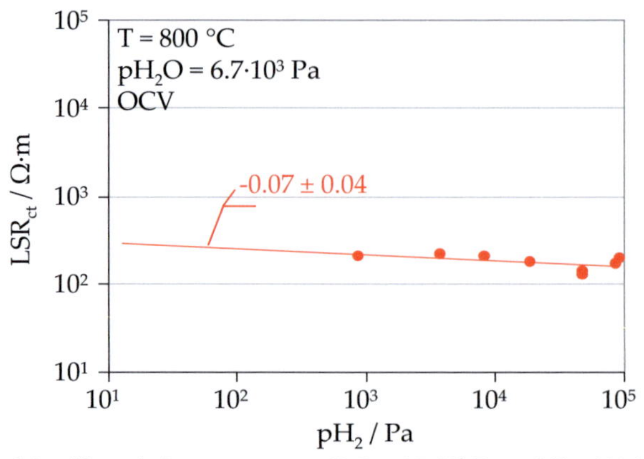

figure 5.2: LSR_{ct} measured for pH_2 variation at constant $pH_2O = 6.7 \cdot 10^3$ Pa and T = 800 °C. N_2 is used as balance, sample: NiYSZ_139.

Compared to the pH_2O dependency, a weaker dependency is observed on pH_2, which is in agreement with literature. The reported dependencies spread from positive values of the exponent a of 0.15 [9] and 0.11 [6] to a stronger negative value of a = -0.26 [7]. Additionally, data by Mizusaki indicates an increased dependency on pH_2 for elevated temperature and low partial pressures [9].

Discussion

In a theoretical study, Skaarup et al. [93] found that the surface coverage of Ni with hydrogen Θ_H changes with gas composition and is non-negligibly small, e.g. for T = 700 °C and $pH_2 = 6.5 \cdot 10^4$ Pa, a value of $\Theta_H = 0.51$ is reported. This may explain the differences in gas composition dependencies reported in the different literature studies.

The strong decrease in LSR_{ct} with increasing pH_2O is counter-intuitive as water is the reaction product. According to the principle of Le Chatelier, an increase in product concentration should decrease the reaction rate. However, the oxidation of H_2 is an electrochemical process and therefore the reaction rates are also strongly affected by the electrode potential. When changing the gas composition, the electrode potential changes as well. In an elementary kinetic modeling study, Bessler at al. attributed the strong dependency to so-called equilibrium potential effects [13, 56].

A further explanation involves the presence of glassy impurities whose electrical properties are strongly dependent on the gas composition. At very low pH_2O the impurity layer behaves as an electrical insulator, whereas at higher pH_2O the proton conductivity increases due to an uptake of water [66]. An estimation of the proton conductivity of 10^{-6} to 10^{-5} $S \cdot cm^{-1}$ at 800 °C is given for silica glass in [66], even higher values are expected for metal ion containing glasses in wet atmospheres. Additionally, the formation of highly volatile $Si(OH)_4$ at high water vapor contents with consequent depletion of the impurity species SiO_2 from the surface could explain the decrease in LSR_{ct} with increasing pH_2O [63].

While the reasoning with a changing electrode potential only explains the observed qualitative trend of LSR upon variation of pH_2O, different sample specific impurity compositions can additionally explain the deviations in absolute LSR values between the different studies in literature. However, as already discussed in chapter 4, the considerable relaxation and degradation mechanisms occurring during electrochemical characterization are considered to strongly account for these deviations as well, especially in the case of pH_2 variation. As previous studies did not specify the conditions of electrochemical characterization and the impurity features, the interpretation of the respective experimental data is questionable.

In modeling studies of the elementary kinetics performed by Bessler and coworkers, the negative proportionality of LSR_{ct} and pH_2O can be reproduced by all applied charge transfer mechanisms (hydrogen, oxygen and hydroxyl spillover) and is caused by the so-called equilibrium-potential effects [16, 56]. In contrast, the pH_2 behavior has been shown to be a distinguishing feature between hydrogen spillover (decrease in LSR for increasing pH_2) and oxygen spillover (increase in LSR for increasing pH_2). With respect to the scatter in experimental data for pH_2 variation and the discrepancies between literature data, a statement on the reaction mechanism is difficult. However, the observed weak dependency is rather ascribable to hydrogen spillover than to oxygen spillover [16, 56].

5.2.2 Temperature

The activation energy E_{act} has been determined in the temperature range 450 °C ≤ T ≤ 800 °C at $pH_2 = 8.44\cdot10^4$ Pa, $pH_2O = 1.69\cdot10^4$ Pa, $pN_2 = 0$ Pa, $\beta_{rel} = 16$ %, see figure 5.3. The trend in LSR_{ct} exhibits a slight kink at 700 °C: a value of $E_{act} = 1.01$ eV is obtained for T ≥ 700 °C, whereas for T ≤ 700 °C the value is higher with $E_{act} = 1.37$ eV.

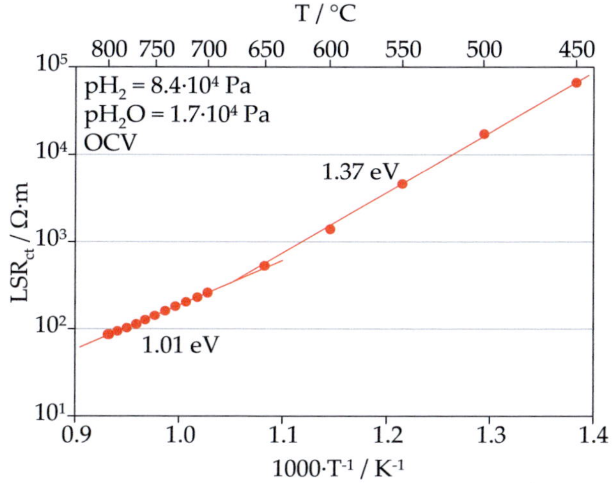

figure 5.3: LSR_{ct} measured for temperature variation at $pH_2 = 8.4\cdot10^4$ Pa, $pH_2O = 1.7\cdot10^4$ Pa ($\beta_{rel} = 16.8$ %) at OCV, sample: NiYSZ_141. The temperature dependency exhibits a kink at 700 °C, yielding two different values for E_{act}.

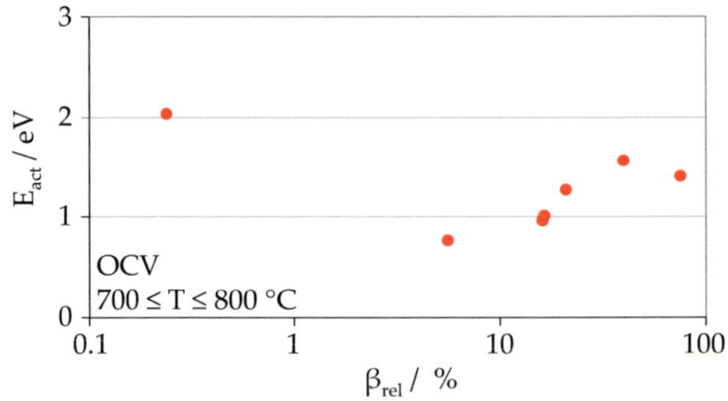

figure 5.4: Activation energy E_{act} for different relative water vapor contents measured between 700 – 800 °C at OCV, sample NiYSZ_141. E_{act} varies considerably with water vapor content, with very high activation energy for dry gas atmosphere.

Additionally, the activation energy has been determined for different gas compositions of pH_2, pH_2O and pN_2 between 700 °C ≤ T ≤ 800 °C. Different values have been obtained for the different gas compositions, see figure 5.4: (i) at very dry atmosphere, an unexpectedly high value of E_{act} = 2.0 eV was obtained, (ii) between 1 – 20 % relative water vapor content it is 0.8 eV ≤ E_{act} ≤ 1.3 eV, and (iii) for β_{rel} > 20 % again slightly higher values of 1.4 – 1.6 eV are observed.

Discussion

In literature, lower values for the activation energy are reported between 0.73 eV at β_{rel} = 4.4 % and 0.96 eV at β_{rel} = 54.4 % by Mizusaki [9] and 0.88 eV at β_{rel} = 0.3 % by Bieberle [6]. De Boer reported a high activation energy of 1.6 eV at β_{rel} = 2.3 % [7].

The activation energies determined in this work refer exclusively to the contribution of charge transfer, as the contributions of gas diffusion and counter electrode have been separated during data analysis, see section 3.3. Thus, the lower activation energies reported in literature can potentially be attributed to a contribution of gas diffusion that is known to have a smaller and even slightly negative activation energy [33]. According to the gas composition dependency of gas diffusion in equation 3.5, the contribution of gas diffusion will be most pronounced at very low and very high water vapor contents.

A comparison to data from Ni/8YSZ cermet anodes is not straightforward, as the thermal behavior of the activation polarization depends on two coupled processes, namely oxygen ion transport in the YSZ matrix and charge transfer at the TPB. Using a transmission line model, Sonn [35] was able to determine separate values for the activation energy of the ionic transport losses (0.90 eV) and the charge transfer resistance (1.35 eV). Thus, a very good agreement for the charge transfer resistance is obtained between 1.35 eV from Sonn and this work with 1.37 eV for 450 ≤ T ≤ 700 °C.

The observed change in activation energy with gas composition can be an indication for a change in reaction mechanism, e.g. an adsorption/desorption or surface transport reaction being rate determining at very dry gas atmosphere. For instance, activation energies between 0.34 - 4.87 eV have been reported by Zhu et al. for typical heterogeneous reaction mechanisms at Ni electrodes [60]; some reactions are even not temperature activated at all. Reaction mechanisms involving impurity species at the TPB are suspected to increase the spread in activation energy.

In a related study, a sensitivity analysis was performed for the elementary kinetic modeling framework that includes heterogeneous chemistry, surface transport and two hydrogen spill-over charge transfer mechanisms [13]. The comparison of sensitivity analysis results for two different temperatures revealed a change in the reaction mechanism between 500 – 900 °C: at 500 °C, the only rate-determining reaction is the charge-transfer step of hydrogen spillover from the Ni surface to hydroxyl ions on the YSZ surface. At 900 °C, this reaction becomes co-limited by a second charge-transfer step (H spillover from the Ni surface to oxygen ions on the YSZ surface) together with hydroxyl diffusion on YSZ, and H_2O desorption from YSZ. This modeling study was able to reproduce the experimentally determined temperature behavior of figure 5.3, including the kink at 700 °C [13].

Similar to the discussion for the dependency of LSR_{ct} on pH_2O, degradation and relaxation processes may be the reason for the discrepancies in literature: for a recording with increasing temperature, the actual temperature activation of LSR_{ct} is superimposed by morphological changes unless the stabilization of the Ni film morphology has been ensured by an initial heat treatment at the maximum temperature.

5.2.3 Polarization Voltage

In order to determine the charge transfer coefficient α, impedance measurements need to be performed under polarization. This procedure allows to determine the resistance of the activation polarization ($R_{ct} = R_{A\beta}$) separated from the contributions of gas diffusion and counter electrode. In contrast, a current-voltage curve does not allow for this separation and therefore the fitting of a Tafel slope to the current-voltage curve yields an erroneous value of α.

The characterization was carried out in potentiostatic mode with stepwise increase of the anodic overpotential (η = 0, 50, 100, 150, 200, 250, 300 mV), while the corresponding current density j was recorded ($pH_2 = 8.3\cdot10^3$ Pa, $pH_2O = 6.6\cdot10^3$ Pa, $pN_2 = 8.6\cdot10^4$ Pa, β_{rel} = 44 %, for T = 700, 750, 800 °C). The differential resistance attributed to charge transfer $dR_{ct}(j)$ is shown in figure 5.5a. It has been derived from the Complex Nonlinear Least Square (CNLS) fits according to the equivalent circuit described in section 3.3. Note that $R_{ct} \equiv R_{A\beta} \equiv dR_{ct}(j = 0)$ in the special case of open circuit voltage.

The Butler-Volmer equation is an implicit equation of current density j as a function of overvoltage η_{ct}. Thus, a graph with η_{ct} as abscissa and j as ordinate is required in order to fit the charge transfer coefficient α. To this end, the differential resistance $dR_{ct}(j)$ is integrated with respect to the electrical current density j, using a linear interpolation and stepwise integration, see equations 5.6 and 5.7:

$$dR_{ct}(j) = \frac{d\eta_{ct}}{dj} \quad 5.6$$

$$\eta_{ct}(j) = \int_0^j dR_{ct}(j)dj + c \quad 5.7$$

With the boundary condition $\eta_{ct}(j=0) = 0$, the integration constant c is zero. By inserting the current densities into the integration, the corresponding activation overpotential η_{ct} can be calculated for each operating point and the Butler-Volmer equation is fitted to the data $j(\eta_{ct})$, see figure 5.5b. The current - overpotential behavior can very well be described by a Butler-Volmer ansatz. The fit yields a charge transfer coefficient $\alpha = 0.69$ for 800 °C, lower values of 0.65 and 0.61 are obtained for 750 °C and 700 °C respectively.

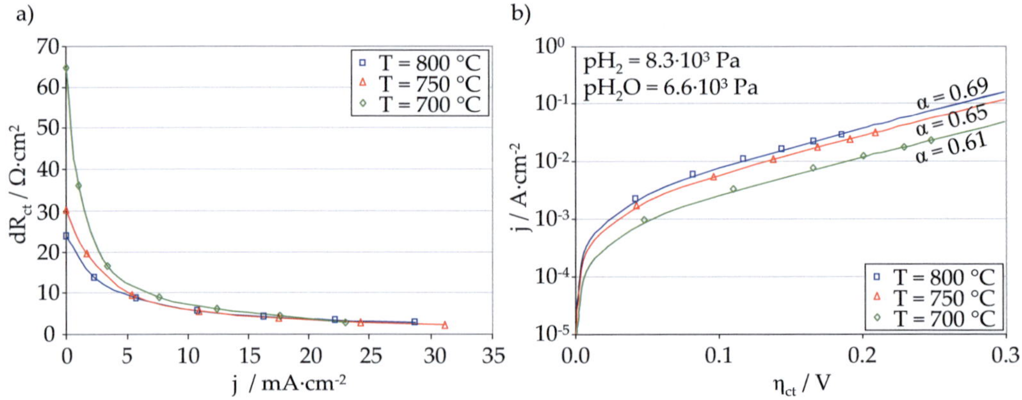

figure 5.5: Determination of charge transfer coefficient α for different temperatures: a) dR_{ct} at various current densities j for polarization voltages of η = 0, 50, 100, 150, 200, 250, 300 mV, and b) current density j as a function of the activation overpotential η_{ct}, together with the respective Butler-Volmer fits ($pH_2 = 8.3\cdot10^3$ Pa, $pH_2O = 6.6\cdot10^3$ Pa, balance N_2, β_{rel} = 44 %, NiYSZ_146)

Discussion

For patterned Ni anodes so far no charge transfer coefficient has been reported following this procedure. The values in the study by de Boers – α_{an} = 0.75 to 1.25 for anodic polarization and α_{cat} = 0.25 to 0.75 for cathodic polarization – are extracted from the current-voltage curve with a fit to the Tafel equation [7]. The same applies for data by Mizusaki with $\alpha_{an} \approx 0.87$ and α_{cat} = 0.36 to 0.53. However, this method is only valid if the gas diffusion losses can be neglected compared to the activation losses. The results in section 3.3 showed that this is not always the case, especially for elevated temperatures with T > 700 °C.

Note that in this work, the charge transfer coefficient α has been calculated according to the Butler-Volmer equation (see equation 5.1) with $n_e = 2$. In several studies, the charge transfer coefficient is determined from the Tafel slope $\alpha'\cdot F/(RT)$ of the logarithmic current density over the polarization voltage, thus assuming $n_e = 1$ [7, 16, 54]. Hence, the resulting value of α' is twice the value of α in this work.

In modeling studies, the charge transfer coefficient is often generally assumed to be 0.5 [94] due to the lack of experimental data. However, this already implies preconditions on the reaction mechanism: from electrochemical theory, a single rate determining charge transfer mechanism is expected to yield a proportionality of $\Delta \ln(j)/\Delta\eta = 0.5 \cdot F/(RT)$, thus $\alpha' = 0.5$; two consecutive charge transfer steps with the second step being rate determining, yield $1.5 \cdot F/(RT)$. The cathodic slope is expected to be as high as to yield $\alpha_{an}' + \alpha_{cat}' = 1$ or 2 [16]. In experimental studies, typically values exceeding $\alpha' = 0.5$ (i.e. $\alpha = 0.25$) are obtained, see results reported above. As explanation, Vogler proposes either a coupling to an additional rate determining step (e.g. diffusion) or two consecutive charge transfer steps [16]. From the sum of anodic and cathodic coefficient exceeding 2, Goodwin assumes that two charge-transfer mechanisms proceed in parallel [54].

5.2.4 Triple-Phase-Boundary Length (l_{TPB})

The effect of the TPB length is shown in figure 5.6 by plotting R_{ct} as a function of $l_{TPB,corr}$. The sample to sample scatter accounts for a deviation in $1/R_{ct}$ values of up to a factor of two. However, a decrease in R_{ct} with increasing $l_{TPB,corr}$ is clearly observable.

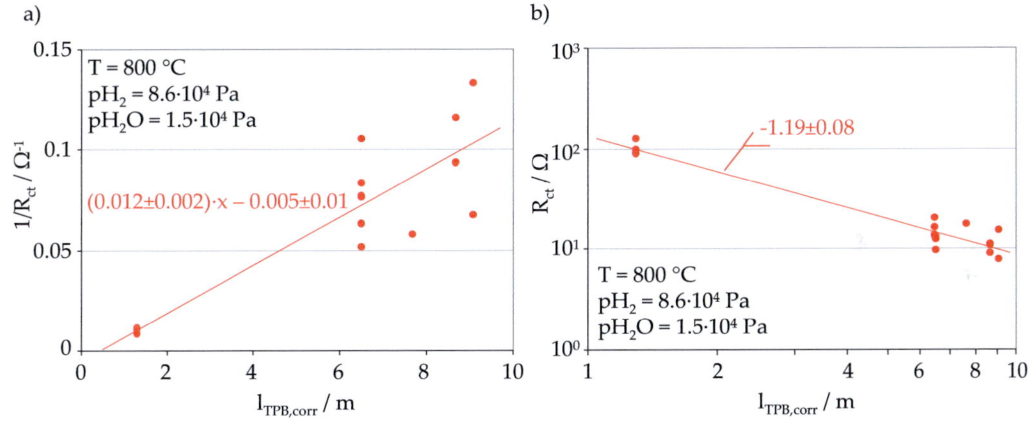

figure 5.6: Polarization resistance of charge transfer reaction R_{ct} for different TPB lengths recorded at similar conditions after stabilization of R_{ct} during thermal exposure ($pH_2 = 8.6 \cdot 10^4$ Pa, $pH_2O = 1.5 \cdot 10^4$ Pa, T = 800 °C): a) proportionality of $1/R_{ct}$ and $l_{TPB,corr}$, and b) double logarithmic graph of R_{ct} as a function of $l_{TPB,corr}$.

Discussion

A linear proportionality between charge transfer resistance and TPB length is expected, as charge transfer is supposed to takes place at the TPB. This is also one of the reasons for the electrochemical characterization with model anodes and the definition of a line specific resistance in the first place. The same trend has been reported by most groups for patterned Ni anodes [5, 7, 9, 95]. However, some were not able to determine the l_{TPB} due to instabilities of the Ni film [27, 29].

Straightforward from the definition of LSR in equation 1.5, a slope of -1 is expected for R_{ct} as a function of l_{TPB} in a graph with double logarithmic scale, see equation 5.8. As can be seen in figure 5.6b, in this work a slightly stronger slope of -1.19 was obtained. Bieberle reported a much weaker slope of -0.7, that is independent on the applied polarization voltage [6]. Values of -0.5 and -0.68 are extracted from the data by Mizusaki [9] and de Boer [7] respectively.

$$\log(R_{ct}) = -\log(l_{TPB}) + \log(LSR_{ct}) + \log(A) \qquad 5.8$$

A comparison of the slopes in the linear graph such as performed in [6] is not adequate: in this case, the proportionality factor corresponds to the reciprocal value of LSR_{ct} and is therefore strongly dependent on gas composition, temperature and polarization voltage, see equation 5.9. Hence, unless the comparison is performed with data at similar measurement conditions, strong deviations in the slope will be obtained.

$$\frac{1}{R_{ct}} = \frac{1}{LSR_{ct}(T, pH_2, pH_2O, \eta)} \cdot l_{TPB} \cdot A \qquad 5.9$$

In the linear graph of $1/R_{ct}$ as a function of $l_{TPB,corr}$, the passage of the fitted line comes close to the axis origin, affirming that the rate determining reaction takes place at the TPB, see figure 5.6a. Furthermore, it shows that in this work the Ni layer was very dense and did not expose a significant amount of TPB other than the l_{TPB} of the pattern design.

5.3 Comparison to Literature Data

For the validation of modeling studies with experimental results, a presentation of the LSR_{ct} values as a function of pH_2O, pH_2 and temperature is desirable [16]. Additionally, such a presentation of LSR_{ct} values is advantageous for the comparison of LSR data from different studies. Based on the weak dependency of LSR_{ct} on pH_2, a plot showing LSR_{ct} as a function of pH_2O has been chosen. Data recorded at varying pH_2 values is calculated to a standard $pH_2 = 10^4$ Pa, considering the obtained pH_2 dependencies described by the exponent $-0.15 \leq a \leq 0.14$, see figure 5.7. This calculation is based on the assumption of linearly independent variables pH_2 and pH_2O, on which also the Butler-Volmer equation with its power law ansatz for the exchange current density j_0 is based, see equations 5.1 and 5.2 [91, 92]. Hence, the following equation applies. (Note that neglecting the dependency of LSR_{ct} on pH_2 (i.e. $a = 0$) would result in a maximum error in LSR_{ct} of 10 %):

$$\log\left[LSR_{ct}(pH_{2,s})\right] = -a \cdot \left[\log(pH_{2,s}) - \log(pH_{2,m})\right] + \log\left[LSR_{ct}(pH_{2,m})\right] \qquad 5.10$$

Here, subscript s represents the standard pH_2 values (in this case 10^4 Pa) and subscript m the pH_2 value of the actual measurement.

The resulting plot for data recorded at 800 °C is shown in figure 5.7. It includes the data recalculated from figure 5.1 (closed symbols) as well as further data points from various measurements at 800 °C. Even though the degradation and relaxation processes described above have been considered in data recording and analysis, considerable deviations in absolute LSR_{ct} values persist. Therefore, under the assumption of a normal distribution of the data points, a 3σ confidence interval has been introduced and is indicated by the colored area in figure 5.7.

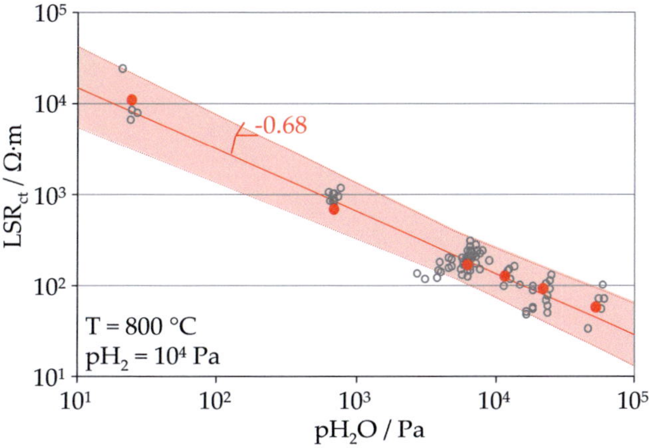

figure 5.7: LSR_{ct} at T = 800 °C calculated for $pH_2 = 10^4$ Pa for data points from figure 5.1 (closed symbol) and further measurement results of this work (open symbol). The calculated confidence interval (3σ) is displayed by the colored area.

Since different studies report data at different temperatures, the experimental data has been extrapolated in order to extend the temperature range. This extrapolation is especially interesting for T > 800 °C where the structural stability of the patterned Ni anode used in this work is insecure (see section 3.1.3). In a first step, the confidence interval has been extrapolated to T = 700 °C using the activation energy (E_{act} = 1.01 eV), see figure 5.8a. The extrapolated confidence interval is in good agreement with results from the electrochemical characterization at 700 °C with the same patterned Ni anodes. Hence, it is concluded that the extrapolation to further temperatures using the activation energy is possible under the assumption of a temperature invariant reaction mechanism. Additionally, accounting for deviations in the activation energy, an uncertainty is introduced to the extrapolation, see figure 5.8a and c.

In order to allow a comparison of all patterned Ni anode data, literature data from Mizusaki, de Boer and Bieberle [6-9] are also included in these graphs. To this end, the reported R_{pol} values as well as the reported TPB length and electrode area are extracted from the respective studies to calculate LSR, see equation 1.5. Note that in this case a separation of charge transfer from gas diffusion is not possible, hence the notation LSR instead of LSR_{ct}. In a next step, LSR values are calculated to $pH_2 = 10^4$ Pa using the parameter dependency a, which is reported in literature, see section 2, table 2.2. The LSR values obtained at the temperature of characterization are shown in the same plot with the LSR_{ct} values and the confidence interval determined in this work. Following this approach, it is possible to compare the results reported in this work with patterned Ni anode literature [6-9], both with respect to absolute LSR values and parameter dependencies.

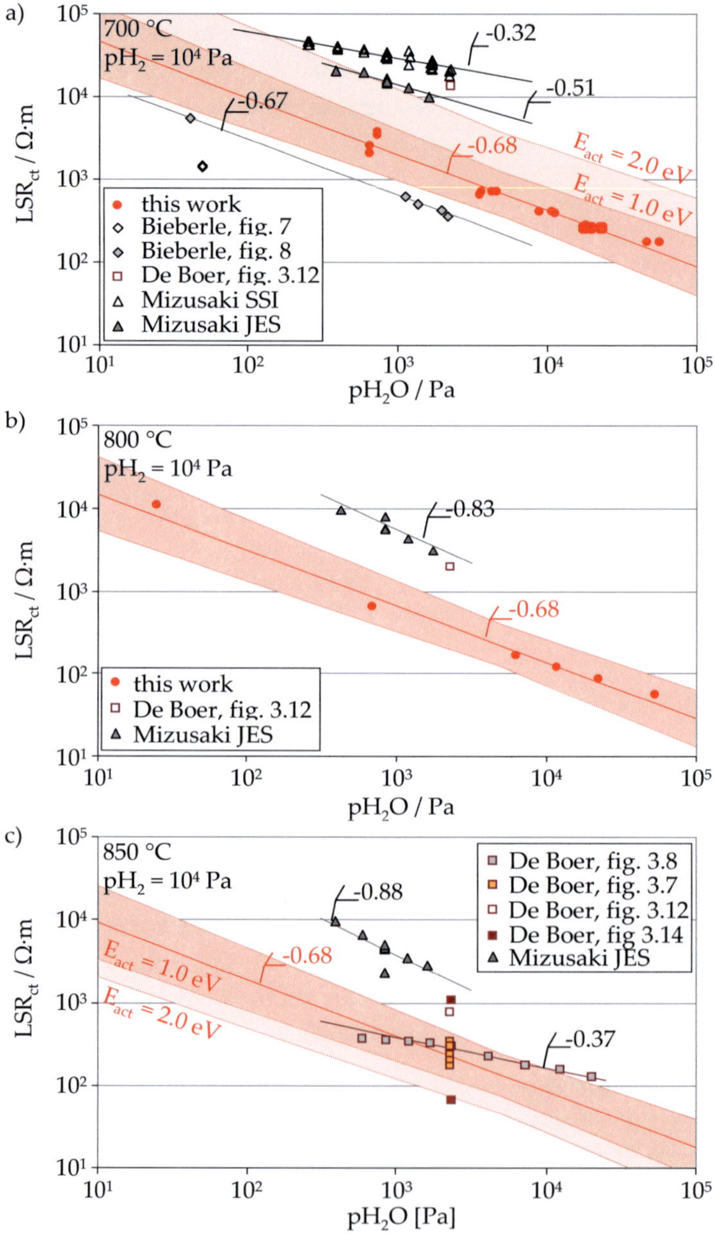

figure 5.8: Comparison of measured LSR_{ct} data with the calculated confidence interval and different literature data [6-9] for $pH_2 = 10^4$ Pa: a) at 700 °C, b) at 800 °C, and c) the extrapolated values at 850 °C. The lighter areas indicate the uncertainty of the activation energy (E_{act} = 2.0 eV instead of 1.0 eV).

LSR values of Mizusaki are high up by one order of magnitude and scatter in pH_2O dependency with values for b between 0.32 and 0.88. LSR values of Bieberle are below, and rather close to the confidence interval determined in this work. Additionally, the pH_2O dependency with b = 0.67 is very similar. However, LSR values at 700 °C differ by almost two orders of magnitude between Mizusaki and Bieberle. Data from de Boer are within the confidence interval at 850 °C, but deviate strongly in pH_2O dependency with b = 0.37. This demonstrates the strong deviations in existing literature on patterned Ni anodes.

In figure 5.9 the LSR values are additionally plotted as a function of temperature. With respect to the different gas compositions applied, data is again calculated for the standard gas composition of $pH_2 = 10^4$ Pa and $pH_2O = 10^3$ Pa. Large deviations persist both for (i) absolute LSR values and (ii) the activation energy with values between 0.75 – 1.6 eV. The deviations in absolute LSR values increase for decreasing temperature and at 600 °C already exceed two orders of magnitude between de Boer and Bieberle.

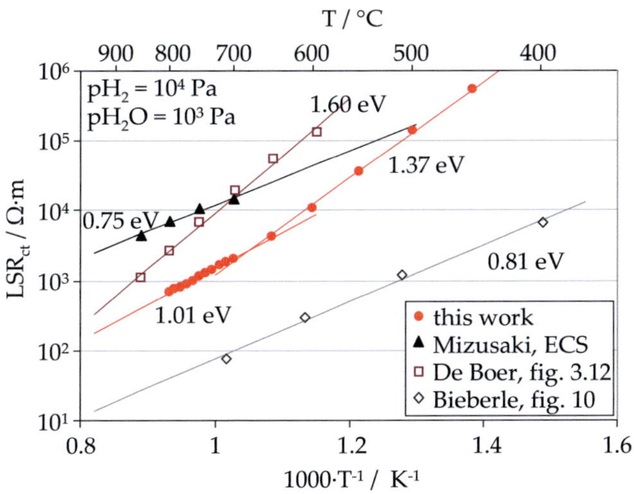

figure 5.9: Comparison of LSR_{ct} data from this work and literature [6-9] as a function of temperature. Data are calculated to the standard gas composition of $pH_2 = 10^4$ Pa and $pH_2O = 10^3$ Pa.

Discussion

The calculation of LSR for different gas compositions and temperatures is based on the assumption of (i) linearly independent parameters pH_2, pH_2O and T and (ii) a constant reaction mechanism over the entire characterized parameter range. The former assumption is at the basis of the Butler-Volmer equation and can be considered as valid, whereas the latter is uncertain with respect to the experimental observation of a change in activation energy for $T \approx 700$ °C (figure 5.3), and for different gas compositions (figure 5.4). However, only with these calculations comparability between the different studies could be obtained.

There is a variety of explanations for the deviations in absolute LSR values: (i) an erroneous estimation of the active TPB length due to delaminated stripes, due to a blocking of the TPB by impurities or due to morphological changes in l_{TPB}, (ii) differences in data recording with respect to degradation rates, or (iii) measurement after parameter variation before the equilibrium state has been attained. De Boer showed that the time of measurement affects the LSR value [7]: for the same measurement conditions, LSR values deviate more than one order of magnitude during an interval of 200 h, see figure 5.8c labeled 'fig 3.14'. Similar observations can be made for Mizusaki and Bieberle if data from different graphs are compared. This underlines once more the importance of the consideration of relaxation and degradation effects.

The different parameter dependencies are assumed to be due to (i) disregard of a potential gas diffusion contribution or (ii) different measurement conditions. The weak (even slightly negative) temperature activation of gas diffusion together with the pronounced dependency on water vapor content can lead to a decrease in observed overall activation energy as well as an increase in pH_2O dependency in the case that gas diffusion and charge transfer contribution are not separated. Furthermore, it has been shown in this work, that the direction of gas variation, the measurement interval and the time of recording are important parameters that strongly influence the results. Especially for the variation of pH_2, this was shown to be a critical issue in data recording due to the long time constants of stabilization, see section 4.3.

By the separation of the different contributions and the consideration of the encountered relaxation and degradation processes, it was possible to achieve a new, consistent data set for patterned Ni anodes in H_2-H_2O atmosphere.

5.4 Comparison to Point Anodes

As discussed earlier in section 2.3, so far no detailed impurity analysis has been performed for patterned Ni anodes. However, this is a decisive point in the interpretation of electrochemical performance of model anodes, as impurities influence the reaction mechanism. The only studies that have combined electrochemical characterization and impurity analysis are reported for point anodes. Therefore, in addition to the comparison of the electrochemical performance of different patterned Ni anodes, the performance of patterned and point anodes is compared with respect to the impurity content. To this end, data of point electrodes reported by Jensen, Høgh and Schmidt are used [23, 67, 68, 71].

As most data by Jensen and Høgh is reported for a gas composition of 3 % H_2O in H_2, all other data is corrected to this gas composition. This correction is performed in accordance to the previous section, using equation 5.10 applied for pH_2 and pH_2O. For the patterned Ni anode, the exponents $a = 0.07$ and $b = 0.68$ are used (see section 5.2). For Ni point electrodes, a smaller exponent of $b = 0.27$ is reported for the pH_2O dependency, the pH_2 dependency is considered to be negligible [18].

In figure 5.10, the LSR values are shown as a function of temperature including the error bars and the calculated activation energy, where applicable. A summary of sample purity and impurity features of the respective samples is given in table 5.1.

*table 5.1: Summary of conditions of samples with respect to purity of components and impurity features after electrochemical characterization (conditions are specified in column 'char.'). Note the difference in impurity specification: * impurity contents refer to substrate after tape casting and sintering, ** impurity contents refer to powder.*

Name	Ref.	Electrolyte	Electrode	Char.	Interface	TPB
This work		8.5YSZ Polycrystalline 0.032* w. % SiO_2	99.98 % Ni SiO_2 amount n.a.	400-800 °C H_2-H_2O	YSZ: impurity imprints at form. Ni grain boundaries Height: 3-7 nm Width: 10-30 nm	Impurity striations Height: 8-15 nm Width: 10-30 nm Entire width: 0.1-1.6 µm
Jensen impure	[23, 68]	8YSZ (7.7-7.8 mol%) Polycrystalline 0.003-0.005** w.% SiO_2	99.8 % Ni 0.003 w.% SiO_2 Ø: 0.38 mm	1000 °C H_2-H_2O	Ni: impurity particles (7-8 µm), rich in Si, Al YSZ: hill-and-valley structure, impurity „bands" (> 20 µm)	Impurity ridge (Si, Al) triangular, several peaks Height: up to 1.6 µm Width: 4 µm
Jensen pure	[23, 68]	8YSZ (7.7-7.8 mol%) Polycrystalline 0.003-0.005** w.% SiO_2	99.995 % Ni 0.0003 w.% SiO_2 Ø: 0.5 mm	1000 °C H_2-H_2O	Ni: few impurity particles YSZ: hill-and-valley structure	Impurity ridge Height: 13-20 nm Width: 150-250 nm
Schmidt pure	[71]	8YSZ Polycrystalline 0.0009** w.% SiO_2	99.999 % Ni 0.00002 w.% SiO_2 Ø: 0.5 mm	850 °C H_2-H_2O	YSZ: hill-and-valley structure, distinction of rather pure rim zone and less pure core zone	Impurity ridge Height: 20 nm
Høgh, 13YSZ	[67]	13YSZ Single crystal, [100] Purity: 99.99*%	99.999 % Ni 0.00002 w.% SiO_2 Ø: 0.5 mm	550-750 °C H_2-H_2O	YSZ: hill-and-valley structure	Impurity ridge Height: 2-16 nm
Høgh, 10YSZ	[67]	10YSZ Single crystal Purity n.a.	99.999 % Ni 0.00002 w.% SiO_2 Ø: 0.5 mm	400-500 °C H_2-H_2O	YSZ: hill-and-valley structure	Impurity ridge Height: 2-16 nm

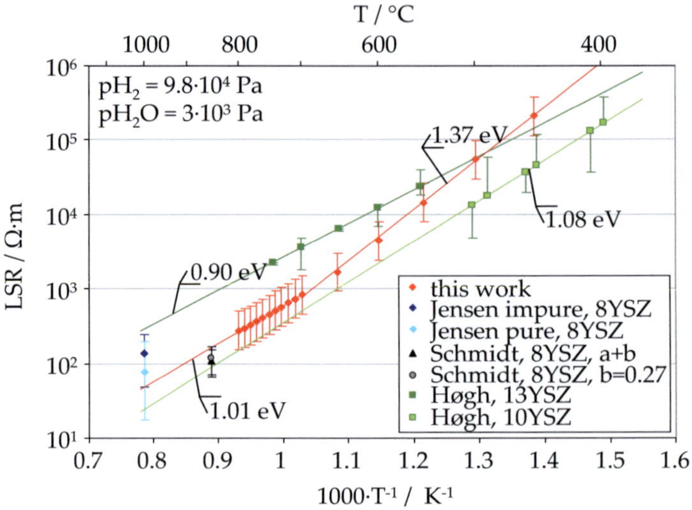

figure 5.10: Comparison of LSR values reported in different studies for Ni point anodes (Jensen [23, 68], Schmidt [71] and Høgh [67]) and patterned Ni anodes of this work at $pH_2O = 3{\cdot}10^3$ Pa and $pH_2 = 9.8{\cdot}10^4$ Pa as a function of temperature. Data from this work is corrected for the gas composition using $LSR = c{\cdot}pH_2^{-a}{\cdot}pH_2O^{-b}$ with a = 0.07 and b = 0.68. As no gas composition dependency is given for data by Schmidt, two corrections have been applied: correction for pH_2 and pH_2O with a = 0.07 and b = 0.68 (denoted 'a+b') and correction only for pH_2O with b = 0.27 determined with data from Høgh (denoted 'b=0.27'). For details to the impurity content see table 5.1

Discussion

The size of the impurity striations at the TPB with a height of up to 20 nm is comparable to values reported for very pure point anodes, whereas heights exceeding 1 µm were observed for less pure Ni electrodes, see table 5.1. As discussed earlier in section 3.4, this very low impurity level was only achievable by special emphasis on the choice of materials and measurement setup.

The comparison of LSR values obtained from electrochemical characterization of model anodes shows that the deviation is within one order of magnitude. This is a very good agreement, especially with respect to the strong sample to sample variation and the considerable difference in l_{TPB} between few millimeters for point and several meters for patterned anodes. By contrast, the strong deviations in electrochemical performance of different patterned anode studies of almost two orders of magnitude indicate strong differences in impurity level as well as differences in the electrochemical characterization with respect to the dynamic processes observed for patterned Ni anodes and point anodes [58, 65, 72].

5.5 Double Layer Capacity

For patterned Ni anodes the electrode-electrolyte interface area (also denoted Two-Phase-Boundary, 2PB) is very large, compared to the small contact area in technical Ni/8YSZ cermet anodes: for the latter the A_{2PB}/l_{TPB} ratio is around 10^{-7} m, compared to 10^{-5} m in the case of a typical patterned Ni anode with 20 µm stripe width. Hence, besides the capacitive contribution of charge transfer (C_{ct}), a considerable capacitive contribution of the electrical double layer (C_{dl}) is expected. The origin of the electrical double layer is a physical charge separation at the interface, leading to the formation of space-charge layers.

As charge transfer takes place at the TBP and the double layer forms at the 2PB, a Line Specific Capacitance (LSC) for the charge transfer contribution and an Area Specific Capacitance (ASC) for the double layer contribution are expected. The overall, experimentally measured capacitance is then the sum of both contributions multiplied with their characteristic measure (l_{TPB} in $m \cdot cm^{-2}$ and A_{2PB} in $cm^2 \cdot cm^{-2}$ respectively):

$$C = C_{ct} + C_{dl} = LSC \cdot l_{TPB} \cdot A + ASC \cdot A_{2PB} \cdot A \qquad 5.11$$

5.5.1 Experimental Determination

By the characterization of patterned Ni anodes with different A_{2PB}/l_{TPB} ratios, the parameters LSC and ASC can be determined experimentally using equation 5.11. Therefore, in this work a sample with a continuous Ni layer of 800 nm thickness and an area of 1 x 1 cm^2 was characterized in addition to typical patterned anodes. The results of temperature variation for LSR_{ct} and C for two samples are shown in figure 5.11.

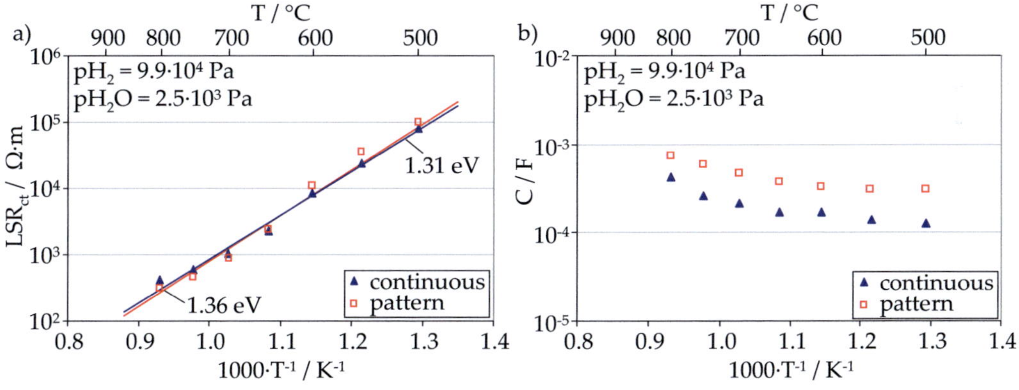

figure 5.11: Electrochemical performance as a function of temperature for two different anode designs: continuous 1 cm^2 Ni surface with 800 nm thickness denoted 'continuous' (sample: DLC_1) and standard patterned Ni anode denoted 'pattern' (sample: NiYSZ_220): a) comparison of LSR_{ct}, and b) comparison of capacitance.

As expected, the values of LSR_{ct}, which consider the respective l_{TPB} of the two sample designs, are in good agreement and thus justify the comparison of the capacitive behavior. The absolute capacitance values differ, with higher values for the patterned anode than for the continuous layer. As two contributions with different causes (l_{TPB} and A_{2PB}) are combined, no single standardized comparison value such as the LSR can be calculated. However, ASC and LSC can be calculated based on at least two data sets with different design, see equation 5.11. The resulting values for variation of temperature and variation of relative water vapor content are shown in figure 5.12.

ASC and LSC both increase with increasing temperature, see figure 5.12a. For ASC this is related to an increase in oxygen ion mobility that is responsible for a double layer capacitance in the first place. Additionally, the dielectric constant of the electrolyte was shown to increase with temperature [40]. LSC increases due to the accelerated charge transfer reaction, see the temperature activation of LSR_{ct} in figure 5.11a.

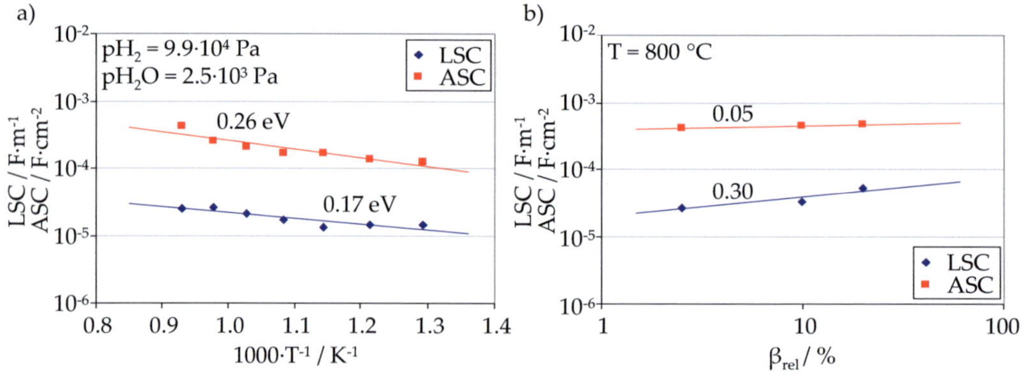

figure 5.12: ASC and LSC calculated from the capacitance of samples with different A_{2PB}/l_{TPB} ratio according to equation 5.11: a) for temperature variation at $pH_2 = 9.9\cdot10^4$ Pa, $pH_2O = 2.5\cdot10^3$ Pa calculated from the data set of figure 5.11b, and b) for variation of the relative water vapor content β_{rel} at 800 °C.

With respect to the gas composition dependency, ASC is supposed to be invariant to gas composition changes. This is in accordance with the experiments, where only a slight increase was observed, see figure 5.12b. By contrast, the charge transfer related LSC increases as expected with increasing water vapor content, since LSR_{ct} decreases due to the accelerated charge transfer reaction.

This procedure of a seperation of the capacitive contributions of charge transfer C_{ct} and double layer C_{dl} reveals a strong difference between the two samples: for the patterned anode, the contribution of the double layer to overall capacitance amounts to 40 – 57 %, whereas it is above 96 % for the continuous Ni electrode. An extrapolation to cermet anode geometry (with $A_{2PB} = 3.8\cdot10^5\ m^2\cdot m^{-3}$, $l_{TPB,v} = 4.0\cdot10^{12}\ m\cdot m^{-3}$ [39]) yields a contribution of double layer capacitance of about 1 %.

This result shows that the double layer needs to be considered in the interpretation of capacitive contribution of patterned Ni anodes. Furthermore, the double layer capacitance represents a distinguishing feature between patterned Ni anodes and Ni/8YSZ cermet anodes.

5.5.2 Theoretical Calculation

A theoretical estimation of the double layer capacitance has been performed following the theory of Stern with a serial connection of an internal Helmholtz double layer at the interface and an outer diffuse double layer according to Gouy and Chapman [91, 96]. Thus, this calculation describes the double layer more detailed than former studies that estimated the double layer capacitance by a simple parallel plate capacitor [37, 41] or a diffuse double layer [40].

The capacitance of the Helmholtz double layer (C_{Hh}) directly at the interface is calculated using the equation for a parallel plate capacitor with the vacuum permittivity ε_0, the dielectric constant ε_r of YSZ, the capacitor area A and the distance d_{O2-} of the plates, see equation 5.12. The distance of the plates is taken as the radius of O^{2-}, d_{O2-} = 140 pm. In literature, a strong variation of the dielectric constant of YSZ between 1 and 60 has been reported [97]. This results in values for C_{Hh} between 0.063 to 3.79 $F\cdot m^{-2}$.

$$C_{Hh} = \frac{\varepsilon_0 \varepsilon_r A}{d_{O^{2-}}} \tag{5.12}$$

The diffuse double layer capacitance (C_{diff}) can be calculated in accordance to the calculations by Hendriks et al. [40] with the elementary charge e, the vacuum permittivity ε_0, the dielectric constant ε_r of YSZ, the density c_{YSZ} of the YSZ unit cell, the Boltzmann constant k_B, the temperature T, and the molar fraction of oxygen vacancies in the YSZ bulk $[V_O^{\bullet\bullet}]_{Bulk}$:

$$C_{diff} = 4e\sqrt{\frac{\pi \varepsilon_0 \varepsilon_r c_{YSZ}}{k_B T}\left[V_O^{\bullet\bullet}\right]_{Bulk}\left(1-\left[V_O^{\bullet\bullet}\right]_{Bulk}\right)} \tag{5.13}$$

For 8.5 mol% YSZ it is $c_{YSZ} = 2.98\cdot 10^{28}$ m^{-3} and $[V_O^{\bullet\bullet}]_{Bulk}$ = 0.07835. Using ε_r between 1 to 60, this yields values for C_{diff} between 1.28 to 9.98 $F\cdot m^{-2}$ at T = 800 °C.

The serial connection of Helmholtz double layer capacitance and diffuse double layer capacitance results in a double layer capacitance C_{dl} between 0.06 and 2.75 $F\cdot m^{-2}$. The ASC values from experimental data of 1 to 5 $F\cdot m^{-2}$ lie in the upper range of the theoretical double layer capacitance. This variation may be due to the estimate of the dielectric constant which is discussed to differ between the space charge region and the bulk [40] or due to additional chemical capacitances e.g. from redox reactions of impurities such as Mn^{2+}/Mn^{3+} [13, 24, 65].

5.6 Conclusions

Parameter dependencies of LSR_{ct} on pH_2, pH_2O, T and polarization voltage have been determined for an extended parameter range with special attention to distortions induced by relaxation and degradation processes. All relevant parameter dependencies for the electrochemical oxidation of H_2 have been determined: the negative dependency of LSR_{ct} on pH_2O (b = 0.68) confirmed observations from literature and the dependency on pH_2 was found to be very small.

This indicates a hydrogen spillover as the rate determining step. Together with the activation energy of E_{act} = 1.01 - 1.37 eV (T = 450 °C - 800 °C) and the charge transfer coefficient α = 0.61 to 0.69 (T = 700 - 800 °C), a complete data set for the modeling of reaction kinetics of patterned Ni anodes has been established.

The proportionality between LSR_{ct} and l_{TPB} has been shown to be close to the expected value of -1 in double logarithmic plot, affirming the assumption of charge transfer at the TPB.

The calculation of LSR_{ct} data to standard conditions and the extrapolation to temperatures higher than the experimentally performed 800 °C enabled for the first time the comparison of data from different literature studies. The observed considerable deviations between previous studies are attributed to (i) disregard of gas diffusion contribution in analysis of impedance spectra, (ii) differences in the measurement procedure regarding relaxation and degradation processes as well as (iii) erroneous estimation of the active TPB length and possible blocking with impurities.

Good agreement is found between the performances of patterned Ni anodes and pure point electrodes at OCV with a deviation of less than one order of magnitude. Additionally, the impurity features at the TPB exhibit a similar and small size, which emphasizes the high purity of the electrodes.

For the first time, the effect of a double layer capacitance at the electrode-electrolyte interface has been considered in the evaluation of the capacitive contribution of patterned Ni anodes. By the electrochemical characterization of samples with different A_{2PB}/l_{TPB} ratio, the separation into a Line Specific Capacitance (LSC) attributed to charge transfer and an Area Specific Capacitance (ASC) attributed to a double layer was enabled. The experimentally determined values come close to the theoretical double layer capacitance.

6 Electrochemistry of the CO Oxidation

This chapter summarizes the results of the electrochemical characterization of patterned Ni anodes in CO-CO_2 atmosphere.

The analysis of the electrochemical oxidation of CO in CO-CO_2 atmosphere represents a first step towards the operation of SOFCs with reformate gases. Especially the comparison between results from the characterization in H_2-H_2O and CO-CO_2 atmosphere is important with respect to differences in the reaction mechanisms. Therefore, the parameter dependencies for the CO-CO_2 atmosphere need to be determined in the same way as for the H_2-H_2O atmosphere.

The experimental setup is explained in section 6.1. In section 6.2, the parameter dependencies according to the Butler-Volmer equation are given. A comparison to existing literature data is performed in section 6.3, and the comparison to results from H_2-H_2O atmosphere is performed in section 6.4.

6.1 Measurement Setup

For the electrochemical characterization described in this chapter, the sample design was composed of a 500 μm wide frame around the actual stripe pattern on a surface of 9.25 x 9.25 mm^2 (thus electrode surface of 10.25 x 15.25 mm^2). For details to the fabrication procedure and the contacting of the samples see chapter 1. Accordingly, the measurement setup described in the same chapter in figure 3.4b has been chosen. This allows a contacting without Ni mesh on the entire surface, but only on two sides of the frame using a special contact block.

Electrochemical impedance spectra (EIS) were recorded over an extended range of gas compositions that avoid Ni oxidation and carbon formation: $4.0 \cdot 10^2$ Pa $\leq pCO \leq 5.1 \cdot 10^4$ Pa and $9.5 \cdot 10^2$ Pa $\leq pCO_2 \leq 9.2 \cdot 10^4$ Pa, resulting in 8 % $\leq \beta_{rel,CO/CO2} \leq$ 99 %. As shown in section 3.2, figure 3.6b, two series of gas variation of pCO and pCO_2 respectively were realized with up to 9 points per series; N_2 was used as balance gas. In accordance to the relative water vapor content for H_2-H_2O atmosphere, $\beta_{rel,CO/CO2}$ was calculated following $\beta_{rel,CO/CO2} = pCO_2 / (pCO + pCO_2)$; the variation was 8 % $\leq \beta_{rel} \leq$ 99 %. Additionally, temperature variation was performed for 700 °C $\leq T \leq$ 800 °C. The polarization voltage has not been varied in order to avoid carbon formation.

The recording of impedance spectra was performed by a Solartron 1260 FRA, in a frequency range of 100 mHz – 1 MHz. Due to the high impedance of the patterned anodes, voltage stimulus was chosen instead of current stimulus; the amplitude was 10 mV.

6.2 Results of Cell Measurements

In accordance to the procedure for H_2-H_2O atmosphere, the parameter dependencies of the electrochemical characterization in CO-CO_2 atmosphere are described using the Butler-Volmer equation. The exchange current density j_0 is described by a power law ansatz for the gas composition (partial pressures of CO and CO_2 respectively), and an Arrhenius type temperature dependency:

$$j_0 = \gamma (pCO)^c (pCO_2)^d \exp\left(-\frac{E_{act}}{RT}\right) \tag{6.1}$$

The parameter dependencies of LSR_{ct} are determined based on the inverse proportionality of exchange current density and resistance ($LSR_{ct} \sim 1 / j_0$) and are described by the exponents c and d (dependency of LSR_{ct} on pCO and pCO_2 respectively) and E_{act} (temperature dependency), where c_4, c_5 and c_6 are proportionality constants:

$$LSR_{ct} = c_4 (pCO)^{-c} \tag{6.2}$$

$$LSR_{ct} = c_5 (pCO_2)^{-d} \tag{6.3}$$

$$LSR_{ct} = c_6 \exp\left(\frac{E_{act}}{RT}\right) \tag{6.4}$$

In a graph with double-logarithmic scale of pCO and pCO_2 as abscissa and LSR_{ct} as ordinate, the negative slopes correspond to the exponents c and d respectively. The activation energy is obtained by plotting $\ln(LSR_{ct})$ as a function of the inverse temperature $1000 \cdot T^{-1}$.

6.2.1 Partial Pressures of CO and CO_2

The dependencies of LSR_{ct} on pCO_2 and pCO are shown in figure 6.1 and figure 6.2 respectively. A strong negative dependency of LSR_{ct} on pCO_2 is observable for both analyzed pCO values. However, the quantitative dependency is slightly different: at $pCO = 3.9 \cdot 10^4$ Pa an exponent of $d = 0.79$ is obtained, and at $pCO = 1.0 \cdot 10^4$ Pa a weaker dependency with $d = 0.61$ is observed. However, the latter dependency exhibits a slight kink at $pCO_2 = 2 \cdot 10^4$ Pa, with a higher slope ($d = 0.76$) for lower pCO_2 values and an attenuated slope ($d = 0.41$) for higher pCO_2 values.

The dependency of LSR_{ct} on pCO is less pronounced: for $pCO_2 = 5.1 \cdot 10^4$ Pa a value of $c = 0.24$ is obtained, and at $pCO_2 = 2.0 \cdot 10^4$ Pa opposite behaviors are found for low and high pCO values.

For pCO < 10^4 Pa, LSR_{ct} decreases with increasing partial pressure (c = 0.47), whereas for higher pCO values, LSR_{ct} increases with increasing pCO (c = -0.34).

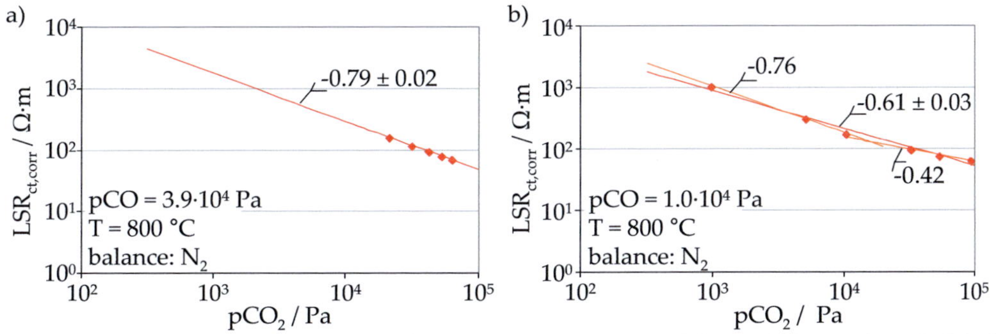

figure 6.1: Parameter dependency of LSR_{ct} as a function of pCO_2 at T = 800 °C for two different values of pCO: a) pCO = $3.9 \cdot 10^4$ Pa, and b) pCO = $1.0 \cdot 10^4$. LSR_{ct} values are corrected for the degradation during recording of parameter variation, yielding $LSR_{ct,corr}$, sample: NiYSZ_291.

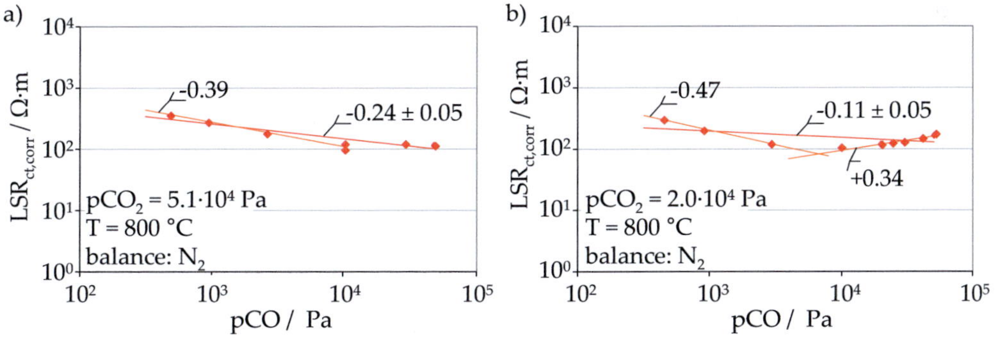

figure 6.2: Parameter dependency of LSR_{ct} as a function of pCO at T = 800 °C for two different values of pCO_2: a) $pCO_2 = 5.1 \cdot 10^4$ Pa, and b) $pCO_2 = 2.0 \cdot 10^4$. LSR_{ct} values are corrected for the degradation during recording of parameter variation, yielding $LSR_{ct,corr}$, sample: NiYSZ_291.

Discussion

The parameter dependencies are similar to those observed for H_2-H_2O, with a strong negative dependency on the reaction product (H_2O, CO_2) and a weaker dependency on the reactant (H_2, CO). By contrast, Boulenouar reported similar dependencies for pCO and pCO_2 of about 0.5 in the case of Ni grid electrodes [47]. For technical Ni/8YSZ cermet anodes, Leonide reported apparent gas composition dependencies of c = -0.058 and d = 0.25 [53, 98, 99]. These are the dependencies of the coupled processes of ionic transport in the YSZ matrix and charge transfer at the TPB, that are expected to be weaker than the pure charge transfer dependency, since ionic transport is gas composition invariant. However, this data affirms the observation of a stronger dependency on the reaction product than on the reactant.

6.2.2 Temperature

The activation energy has been determined in the temperature range of 700 - 800 °C. A value of 1.42 eV is obtained for pCO = $3.9 \cdot 10^4$ Pa and $pCO_2 = 4.2 \cdot 10^4$ Pa, see figure 6.3. Additionally, the activation energy in three operation points of pCO variation at $pCO_2 = 2.0 \cdot 10^4$ Pa has been determined, see figure 6.4a. A clear difference in activation energy is discernable for the three gas compositions between 0.85 eV to 1.42 eV, with increasing activation energy for increasing pCO. Additionally, the pCO dependency of LSR_{ct} for T = 700, 750, 800 °C is shown in figure 6.4b. The previously observed kink in dependency of LSR_{ct} on pCO at around pCO = $2 \cdot 10^4$ Pa is reproduced for all temperatures.

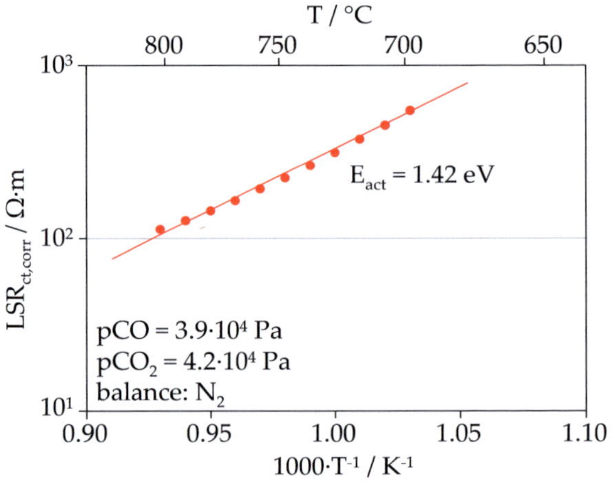

figure 6.3: Temperature dependency of LSR_{ct} recorded at pCO = $3.9 \cdot 10^4$ Pa and $pCO_2 = 4.2 \cdot 10^4$ Pa. The activation energy is E_{act} = 1.42 eV. LSR_{ct} values are corrected for the degradation during recording of parameter variation, yielding $LSR_{ct,corr}$. Sample: NiYSZ_291.

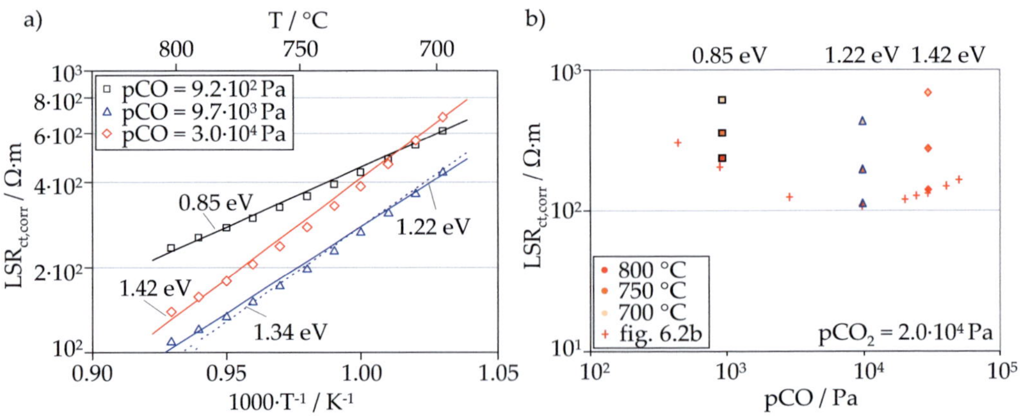

figure 6.4: a) Temperature dependency of LSR_{ct} recorded for different pCO at constant $pCO_2 = 2.0 \cdot 10^4$ Pa. The activation energy varies considerably between E_{act} = 0.85 - 1.42 eV, and b) LSR_{ct} as function of pCO for T = 700, 750, 800 °C together with the data series at 800 °C from figure 6.2b, $pCO_2 = 2.0 \cdot 10^4$ Pa. Values are corrected for the degradation during recording of parameter variation, yielding $LSR_{ct,corr}$. sample: NiYSZ_291.

For Ni grid electrodes, no activation energy has been reported by Boulenouar. For Ni/8YSZ cermet anodes, a value of 1.23 eV is reported. However, this is an apparent parameter dependency, including the contribution of ionic transport and charge transfer.

Discussion

A summary of the dependencies of LSR_{ct} on parameter variations of gas composition and temperature as a function of pCO and pCO_2 is given in figure 6.5. It is noteworthy that both the pCO_2 dependency as well as the pCO dependency change at about pCO = 10^4 Pa, $pCO_2 = 2{\cdot}10^4$ Pa. This is in accordance with the change in activation energy between 0.82 eV for pCO < 10^4 Pa and 1.42 eV for pCO > 10^4 Pa. At pCO = 10^4 Pa, an intermediate activation energy of 1.22 eV is obtained. Together, these reproducible changes in the parameter dependencies are strong indicators for a change in the reaction mechanism within the characterized parameter range. One possible limit for gas compositions with different reaction mechanism is the pCO_2/pCO ratio: as shown in figure 6.5, the line indicating a value of $\beta_{rel,CO/CO2}$ = 60 % separates the two regions with different (i) dependencies on pCO and pCO_2, and (ii) different activation energies. According to the findings in H_2-H_2O atmosphere, the change in mechanism may include a co-limitation of charge transfer, e.g. with adsorption/desorption [13].

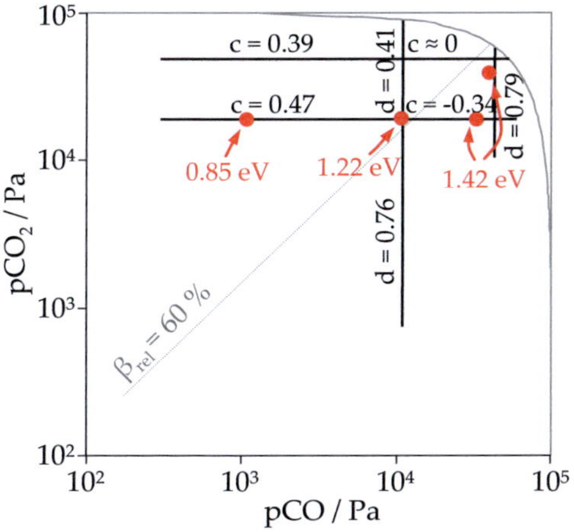

figure 6.5: Summary of the parameter dependencies c, d and E_{act} of LSR_{ct} as a function of pCO and pCO_2 together with a possible limit for a change in the reaction mechanism at $\beta_{rel,CO/CO2}$ = 60 %. The values of c and d were determined at 800 °C.

6.2.3 Triple-Phase-Boundary Length (l_{TPB})

The effect of the TPB length is shown in figure 6.6 by plotting R_{ct} as a function of $l_{TPB,corr}$. In order to determine whether the rate determining reaction step takes place at the TPB, samples with two different values of l_{TPB} have been characterized in CO-CO_2 atmosphere. However, a quantitative analysis would require further values of l_{TPB}.

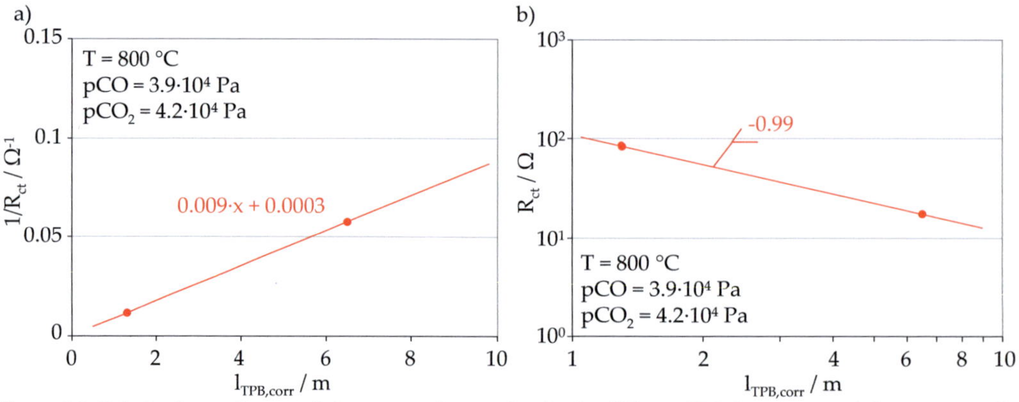

figure 6.6: Polarization resistance of charge transfer reaction R_{ct} for different TPB lengths recorded at same conditions ($pCO = 3.9 \cdot 10^4$ Pa, $pCO_2 = 4.2 \cdot 10^4$ Pa, T = 800 °C): a) proportionality of $1/R_{ct}$ and l_{TPB}, b) double logarithmic graph of R_{ct} as a function of $l_{TPB,corr}$. Values of l_{TPB} have been corrected for the increase due to grain growth, yielding $l_{TPB,corr}$.

In the graph with $1/R_{ct}$ as a function of l_{TPB}, the linear fit of the data points yields an axis intercept with the ordinate of 0.0003 and thus comes very close to the origin, see figure 6.6a. In the double logarithmic plot, a slope of -0.99 is obtained which is very close to the expected -1, see figure 6.6b. Therefore, it can be concluded that the electrochemical oxidation of CO takes place at the TPB and consequently the polarization resistance is governed by a constant LSR.

6.3 Comparison to Literature Data

Sparse work is reported that realized systematic parameter variations of one parameter at a time that allows determining quantitative parameter dependencies. Among them is no study with patterned Ni anodes. Hence, the only source for a comparison of the electrochemical performance of model anodes is the study with Ni grid electrodes by Boulenouar [47]. A direct comparison of LSR data with Ni grid electrodes is difficult since the value of l_{TPB} is not known exactly. Nevertheless, a quantitative comparison is attempted in this work. To that aim, the polarization resistances are extracted from the study by Boulenouar and a coarse estimate of the TPB length (l_{TPB}) is determined: assuming a reaction line at both sides of each Ni wire and a Ni grid with 400 wires per inch, a value of $l_{TPB} = 6.27\ m \cdot cm^{-2}$ is obtained ($A = 0.071\ cm^2$). The lower error bar then corresponds to one reaction line underneath each wire.

A difference persists with respect to the temperature of characterization: whereas data in this work has been recorded at 800 °C, data by Boulenouar is reported at 850 °C and no information is given on the activation energy. Additionally, no correction for the gas composition is performed for this comparison as similar gas compositions were applied for pCO_2 and pCO variation respectively.

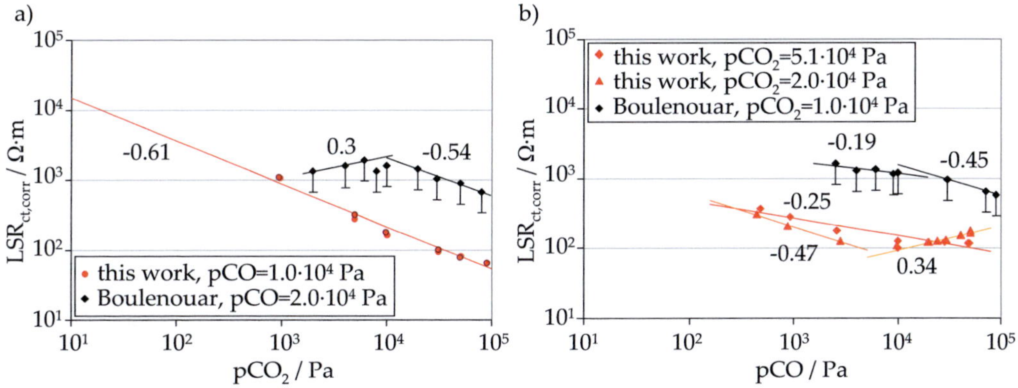

figure 6.7: Comparison of patterned Ni anode results obtained in this work to literature data of a Ni grid electrode [47]: a) for variation of pCO_2, and b) variation of CO. Note that a slight difference in gas composition exists and that data by Boulenouar is reported for 850 °C, contrary to 800 °C in this work. The error bars for the Ni grid electrode indicate an error in l_{TPB} estimation of a factor of 2.

The resulting graphs for pCO_2 and pCO variation are given in figure 6.7. Even though data by Boulenouar is reported for higher temperatures (T = 850 °C), the absolute LSR values exceed data from this work (T = 800 °C) by up to one order of magnitude. Possible causes include (i) poor contacting between Ni grid and electrolyte substrate and thus an overestimation of active l_{TPB} or (ii) a considerable amount of impurities at the TPB and the electrode-electrolyte interface that have segregated from YSZ bulk and Ni grid during the annealing process of the Ni grid on the YSZ substrate at 1400 °C.

Comparing the pCO_2 dependency, the same trend is observed between data from this work (d = 0.61 to 0.79) and Boulenouar (d = 0.54) for $pCO_2 > 10^4$ Pa, whereas at lower pCO_2 values, opposite behavior is reported by Boulenouar with d = -0.3, see figure 6.7a.

Comparing the pCO dependency, no agreement is found, see figure 6.7b. Whereas Boulenouar reported a negative proportionality between LSR_{ct} and pCO throughout the investigated pCO range with c = 0.19 - 0.45, in this work a kink is observed in proportionality for $pCO_2 = 2{\cdot}10^4$ Pa. For low pCO values, negative proportionality (c = 0.47) is observed and direct proportionality at higher pCO values (c = -0.34).

Discussion

Boulenouar et al. analyzed impedance spectra with an equivalent circuit composed of a double layer capacitance in parallel to a series connection of charge transfer resistance and two RC elements [47]. The physical origin of the RC elements is a limitation by adsorption. Boulenouar suggests adsorption of atomic oxygen on the electrode or the electrolyte and adsorption of CO on the electrode. In this work, charge transfer is represented by an RQ element in series to two RQ elements for the contributions of gas diffusion and counter electrode (see section 3.3). No further separation was necessary to represent the impedance spectra. A better agreement between the parameter dependencies on pCO and pCO_2 of both studies is obtained when data by Boulenouar is interpreted as the sum of R_{ct} and the two resistive contributions of the RC elements. In this case, the resulting dependencies are c = 0.25 and d = 0.5.

Furthermore, the observed differences in electrochemical performance between this work and Ni grid electrodes may be explained by a non negligible influence of impurities on the performance of the grid electrode. No information is available on the purity of the electrode and electrolyte materials for preparation and neither SEM images nor further impurity analysis are reported. Especially due to the described annealing process of the Ni grid on the YSZ substrate at 1400 °C, it must be assumed that impurities from the bulk materials have segregated to the surface and the TPB. Impurities at the surface and the TPB may (i) change the kinetics of adsorption/desorption and (ii) slow down the charge transfer reaction at the TPB due to impurity barriers [22, 63]. This may explain the higher absolute LSR values as well as different parameter dependencies reported by Boulenouar compared to the results obtained in this work.

6.4 Comparison to H_2-H_2O Data

In the following, a comparison of the LSR_{ct} values determined for pH_2 and pCO as well as pH_2O and pCO_2 is performed. In figure 6.8, the dependencies of LSR_{ct} on the partial pressures of pCO_2 and pH_2O are given for pCO = pH_2 = 10^4 Pa. Both the absolute LSR_{ct} values as well as the dependency on pCO_2 (d = 0.61) and pH_2O (d = 0.68) are unexpectedly similar.

The variation of pCO and pH_2 for two different values of pCO_2 is shown in figure 6.9. Again, similar LSR_{ct} values are obtained. However, the agreement with respect to parameter dependency is not as good as between pH_2O and pCO_2. For CO-CO_2 atmosphere, a clear and reproducible kink was observed, whereas for pH_2 variation no such pronounced kink was found. Nevertheless it is affirmed that in both atmospheres the dependency on reactant concentration is different from the dependency on product concentration, with a weaker dependency for the reactant than for the product.

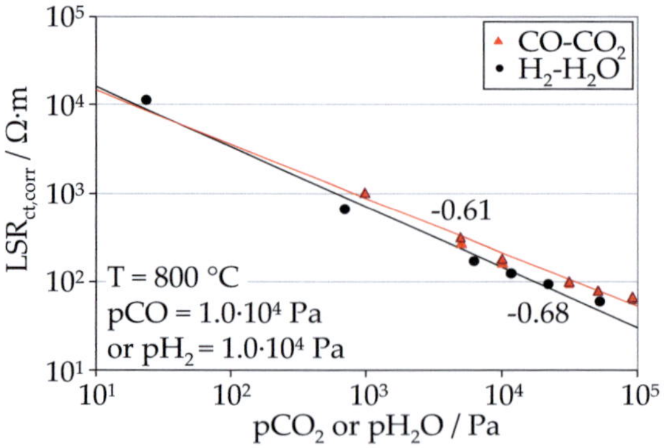

figure 6.8: Comparison of LSR_{ct} values of H_2-H_2O and CO-CO_2 atmosphere for variations of pH_2O and pCO_2 respectively (pH_2 = 1.0·10^4 Pa, pCO = 1.0·10^4 Pa, T = 800 °C). Data originates from figure 5.1 and figure 6.1.

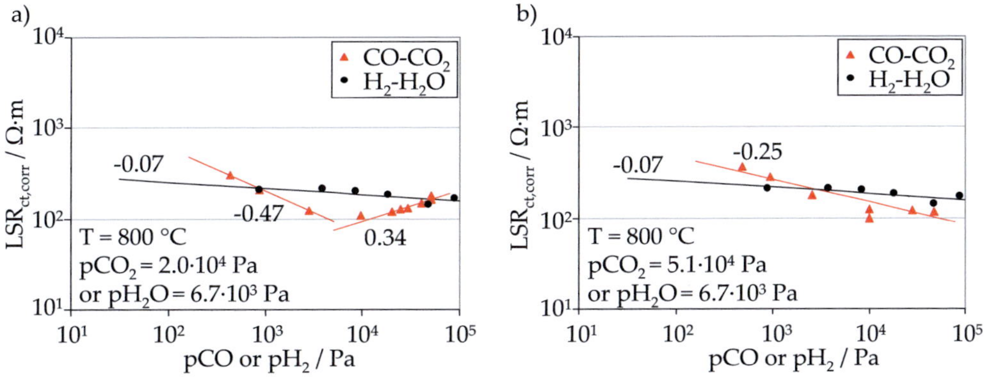

figure 6.9: Comparison of LSR_{ct} values of H_2-H_2O and CO-CO_2 atmosphere for variations of pH_2 and pCO respectively: a) $pH_2O = 6.7 \cdot 10^3$ Pa, $pCO_2 = 2.0 \cdot 10^4$ Pa, T = 800 °C, b) $pH_2O = 6.7 \cdot 10^3$ Pa, $pCO_2 = 5.1 \cdot 10^3$ Pa, T = 800 °C. Data originates from figure 5.2 and figure 6.2.

The activation energy for 700 °C ≤ T ≤ 800 °C for H_2-H_2O at standard conditions is 1.01 eV. Herein, changes in the activation energy for different gas composition were observed between 0.8 to 2.0 eV. The electrochemical characterization in CO-CO_2 atmosphere exhibits a difference in activation energy from 0.85 to 1.42 eV, with increasing activation energy for increasing pCO.

Discussion

The comparison of results from electrochemical characterization in H_2-H_2O and CO-CO_2 gives evidence on the reaction mechanism in both atmospheres. So far, studies advocating the same reaction mechanism [44, 45] as well as completely different mechanisms [14, 29, 46-50] have been reported. The similar characteristics in both atmospheres found in this work indicate similar reaction mechanisms for the electrochemical oxidation of H_2 and CO.

Although a lot of experimental and modeling studies have been performed in H_2-H_2O, no evidence exists on the rate determining mechanisms, yet most results are in favor of a rate limitation by H spillover [13, 16, 39, 54, 100]. In CO-CO_2 atmosphere, a H spillover would correspond to a hop of a Ni-adsorbed CO to the YSZ surface and instantaneous reaction with an oxygen ion to form CO_2. A similar reaction mechanism has been introduced in the study by Lauvstad for the interpretation of experimental data with point anodes [14]. However, results suggest a preferential reaction mechanism with two adsorbed species (O and CO) and charge transfer by O spillover [49]. In a modeling study with five different reaction mechanisms of O spillover mechanisms (one or two step mechanisms), Yurkiv found best agreement for data by Lauvstad [49] with a two step charge transfer (transfer of one electron to the YSZ surface and a consecutive O spillover from YSZ to Ni surface) [61]. However, no CO spillover was included in this modeling framework.

Interstitial transfer of CO has not been discussed so far, as CO diffusion in Ni and YSZ is expected to have no influence due to the large molecule size and the low solubility. With respect to the similar electrode performance, this renders interstitial transfer also improbable for H_2-H_2O. It may contribute in parallel to a surface spillover as proposed in [59], but it is not assumed to account for the entire exchange current density.

However, the observed difference in relaxation and degradation behavior for the characterization in H_2-H_2O and CO-CO_2 atmosphere may be interpreted as an indicator for a difference in reaction mechanisms: long settling times were observed upon variation of pH_2O and especially pH_2, and the degradation rate for LSR_{ct} determined in H_2-H_2O atmosphere is higher than the one in CO-CO_2 atmosphere. It is suggested that the long settling times are caused by a slow equilibration of species concentration between bulk, surface and gas phase in the case of H_2-H_2O atmosphere. Due to the low solubility of CO in Ni and YSZ, no equilibration takes place for CO-CO_2 atmosphere. A limitation of LSR_{ct} involving bulk species as observed for the H_2-H_2O atomsphere may be explained by an interstitial charge transfer.

6.5 Conclusions

Parameter dependencies of LSR_{ct} on pCO, pCO_2 and T have been determined with patterned Ni anodes for a large parameter variation under avoidance of Ni oxidation and carbon formation. The following parameter dependencies for the electrochemical oxidation of CO have been obtained: (i) a negative dependency of LSR_{ct} on pCO_2 (d = 0.61 – 0.79), (ii) a kink in the dependency on pCO at pCO $\approx 2\cdot10^4$ Pa from negative to direct proportionality or weaker dependency, depending on the value of pCO_2, (iii) a change in activation energy from 0.85 – 1.42 eV with increasing pCO. Additionally, the expected proportionality between LSR_{ct} and l_{TPB} has been affirmed. In combination with the results from experimental characterization in H_2-H_2O, these new results for the CO-CO_2 atmosphere represent a powerful data basis e.g. for modeling studies of the reaction mechanisms at SOFC anodes.

A comparison of LSR_{ct} data with data for grid anodes revealed considerably higher values for the grid electrode (deviations exceeding one order of magnitude). It is suggested that these deviations as well as different parameter dependencies are caused by a considerable impurity content as a consequence of an annealing process at 1400 °C.

Similar results for (i) absolute LSR_{ct} values and (ii) parameter dependencies have been obtained for the electrochemical characterization in H_2-H_2O and CO-CO_2 atmosphere, indicating similar reaction mechanisms in both atmospheres. However, so far modeling studies advocate H spillover for H_2-H_2O atmosphere, whereas mostly O spillover mechanisms have been implemented for CO-CO_2 atmosphere. This emphasizes the need of a detailed modeling study for CO-CO_2.

Interstitial charge transfer is assumed ineffective for CO-CO_2, whereas it may contribute to the current density in H_2-H_2O and may be responsible for the observed slow relaxation processes.

7 Comparison to Technical Cermet Anodes

In this chapter, the electrochemical performance obtained for patterned Ni anodes is compared to data reported for technical Ni/8YSZ cermet anodes. This comparison is important with respect to the question, whether findings from model anodes can be transferred to cermet anodes. For this comparison, the coupled processes in a cermet – namely charge transfer at the TPB and ionic and electronic transport processes in electrolyte and electrode – are implemented in a transmission line model.

The transmission line model is introduced in section 7.1. A comparison to results obtained by considering charge transfer exclusively and neglecting transport processes is given in section 7.2. The effect on the parameter dependencies with the denotation of apparent parameter dependencies is shown in section 7.3. Finally, results of patterned and cermet anodes are compared in section 7.4 with respect to parameter variations of pH_2, pH_2O, pCO, pCO_2 and T.

7.1 Transmission Line Model

In technical Ni/8YSZ cermet anodes, a porous mixture of electrode and electrolyte particles is realized in order to obtain a large number of electrochemically active TPB points in the anode volume [1]. This geometry and the high ionic conductivity in the YSZ enable an expansion of the active reaction zone from the electrode-electrolyte interface into the anode volume. The loss mechanisms in such an electrode include contributions of (i) electrochemistry, (ii) gas diffusion and (iii) gas conversion. Herein, the electrochemistry includes charge transfer at the TPB as well as the ionic and electronic transport. It has been shown that the gas phase contributions can be separated from electrochemistry e.g. by the calculation of the Distribution of Relaxation Times (DRT) [33, 35, 83, 90].

A basic equivalent circuit element for the contribution of electrochemistry is the Transmission Line model (TL) [101, 102]. Sonn et al. have applied this model for the interpretation of a screen-printed Ni/8YSZ cermet anode [35].

The scheme of a typical Ni/8YSZ cermet anode with the TPB distributed in the anode volume is given in figure 7.1a. The corresponding transmission line model is composed of (i) the distributed resistances χ_1 and χ_2, that represent the resistance load per unit length (unit: $\Omega \cdot m^{-1}$) in the electrolyte and the electrode respectively, and (ii) the interface impedance ζ at the TPB, modeled as RQ element, see figure 7.1b. Ionic and electronic current density (i_{ion} and i_{el} respectively) as a function of the distance from the electrode-electrolyte interface are shown in figure 7.1c: i_{el} increases with increasing distance from the interface. The penetration depth λ is defined as the distance where 63 % of the current exchange from ionic to electronic has taken place. Hence, the penetration depth is an indicator for the spatial extension of the active reaction zone.

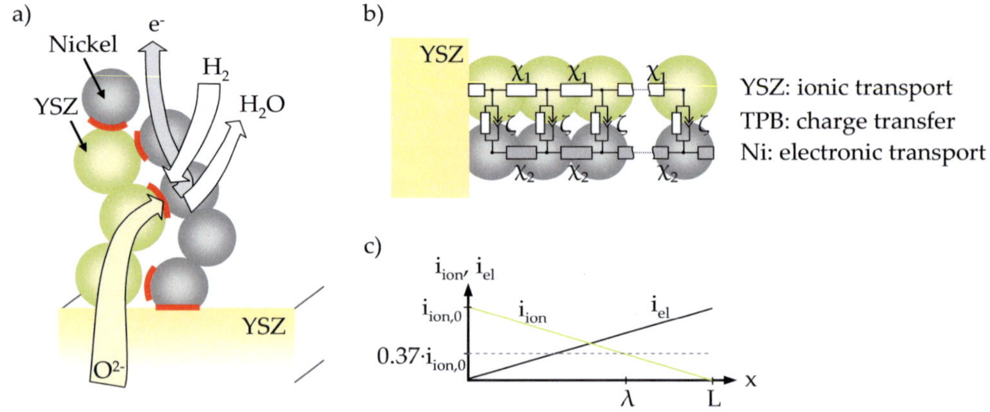

figure 7.1: a) Scheme of transmission line model with location of electrochemical oxidation at the TPB (indicated by the bold lines), b) equivalent circuit of the cermet anode by a transmission line: χ_1 and χ_2 represent ionic and electronic transport, ζ represents the charge transfer at the TPB, c) ionic and electronic current density (i_{ion} and i_{el} respectively) as a function of the distance from the electrode-electrolyte interface. The penetration depth λ indicates the position where 63 % of the current exchange from ionic to electronic has taken place.

The impedance of the transmission line model with a geometric length of L is given by [101]:

$$Z_{TL} = \frac{\chi_1 \chi_2}{\chi_1 + \chi_2}\left(L + \frac{2\lambda}{\sinh(L/\lambda)}\right) + \lambda \frac{\chi_1^2 + \chi_2^2}{\chi_1 + \chi_2} \coth\left(\frac{L}{\lambda}\right) \qquad 7.1$$

Herein, the penetration depth λ is defined by:

$$\lambda = \sqrt{\frac{\zeta}{\chi_1 + \chi_2}} \qquad 7.2$$

Due to the high electrode conductivity, it is $\chi_1 >> \chi_2$. In the case of Ni/8YSZ cermet anodes, the simplification $\chi_2 = 0$ is applied, thus:

$$Z_{TL} = \lambda \cdot \chi_1 \cdot \coth\left(\frac{L}{\lambda}\right) \qquad 7.3$$

$$\lambda = \sqrt{\zeta / \chi_1} \qquad 7.4$$

The ionic transport is represented by a distributed ohmic resistance $z_1 = r_1 = \chi_1$. The total resistance R_1 in the electrode is obtained by integration over the anode thickness L, see equation 7.5. The charge transfer is represented by an impedance length $z_3 = \zeta$ (unit: $\Omega \cdot m$), thus the total resistance of charge transfer Z_3 (in the case of direct current R_3) decreases with increasing anode thickness, see equation 7.6.

$$R_1 = z_1 \cdot L = \chi_1 \cdot L \tag{7.5}$$

$$Z_3 = \frac{z_3}{L} = \frac{\varsigma}{L} \tag{7.6}$$

As charge transfer is represented by an RQ element, it is:

$$z_3 = \frac{r_3}{1 + r_3 q_3 \cdot (i\omega)^n} \tag{7.7}$$

In the case of direct current, it is $z_3 = r_3$ und thus the resistance of the transmission line model R_{TL} is calculated by:

$$R_{TL} = \sqrt{r_1 r_3} \cdot \coth\left(L \sqrt{\frac{r_1}{r_3}} \right) \tag{7.8}$$

Herein, the quantity r_1 is calculated based on the effective ionic conductivity of the electrolyte $\sigma_{ion,eff}$ and the electrode area A:

$$r_1 = \frac{1}{\sigma_{ion,eff} \cdot A} \tag{7.9}$$

And the quantity r_3 is calculated based on the line specific resistance of charge transfer LSR_{ct}, the TPB length per volume $l_{TPB,v}$ and the electrode area A:

$$r_3 = \frac{LSR_{ct}}{l_{TPB,v} \cdot A} \tag{7.10}$$

An overview of the components of the transmission line model and their definitions are given in table 7.1.

table 7.1: Overview of the components in the transmission line model.

Value	Process	Unit	Distributed impedance	Distr. DC resistance	Total DC resistance	Value
χ_1	Ionic transport in YSZ	$\Omega \cdot m^{-1}$	$z_1 = r_1$	r_1	$R_1 = r_1 \cdot L$	$r_1 = \frac{1}{\sigma_{ion,eff} \cdot A}$
χ_2	Electronic transport in Ni	$\Omega \cdot m^{-1}$	$z_2 = r_2$	r_2	$R_2 = r_2 \cdot L$	$r_2 = \frac{1}{\sigma_{el,eff} \cdot A} \approx 0$
ζ	Charge transfer at TPB	$\Omega \cdot m$	$z_3 = \frac{r_3}{1 + r_3 q_3 \cdot (i\omega)^n}$	r_3	$R_3 = \frac{r_3}{L}$	$r_3 = \frac{LSR_{ct}}{l_{TPB,v} \cdot A}$

The case distinctions of (i) very high charge transfer resistance and (ii) very low ionic conductivity are summarized in table 7.2. In the former case ($R_3 >> R_1$), the resistance of the transmission line model is almost equal to R_3 and the expansion of the active reaction zone into the electrode volume is high. This case is denoted charge transfer limitation. In the latter case (transport limitation), the penetration depth is small.

table 7.2: Case distinction of high charge transfer resistance R_3 and low ionic conductivity (high R_1) respectively, together with their implication on the transmission line resistance R_{TL} and the penetration depth λ.

Case	Description	R_{TL}	Pen. depth λ
$R_3 >> R_1$	Charge transfer limitation	$R_{TL} = \frac{1}{3}R_1 + R_3$	$\lambda = \sqrt{\frac{r_3}{r_1}} \nearrow$
$R_1 >> R_3$	Transport limitation	$R_{TL} = \sqrt{R_1 R_3}$	$\lambda = \sqrt{\frac{r_3}{r_1}} \searrow$

7.2 Line Specific Resistance versus Transmission Line Impedance

The quantities of Line Specific Resistance (LSR) and Area Specific Resistance (ASR) have been introduced to enable the comparison of the electrochemical performance of samples with different geometry (either l_{TPB} or electrode area A). The former is common in the analysis of model anode performance, the latter for cermet anodes. In principle, both quantities are interdependent by the parameter l_{TPB}, see equation 7.11. However, the calculation of ASR_{ct} values based on LSR_{ct} and l_{TPB} is only valid if charge transfer exclusively contributes to the polarization resistance, while ionic transport is neglected. In the case of cermet anodes this does not apply. Correspondingly, the comparison of the performance of cermet and patterned anodes by the calculation of LSR_{ct} following equation 7.11 is not possible.

$$LSR_{ct} = R_{ct} \cdot A \cdot l_{TPB} = ASR_{ct} \cdot l_{TPB} \qquad 7.11$$

The error introduced by neglecting the ionic transport for the calculation of ASR is demonstrated in figure 7.2 by the comparison of ASR_{ct} and ASR_{TL} as a function of pH_2O. Here, ASR_{ct} is the quotient of LSR_{ct} and l_{TPB}, while ASR_{TL} is calculated according to equations 7.8 to 7.10 and takes into account the losses due to ionic transport. Therein, the ionic conductivity is assumed to be independent on gas composition. The deviation between ASR_{ct} and ASR_{TL} is a function of pH_2O. With decreasing penetration depth and thus a limitation by ionic transport, the contribution of ionic transport to the polarization resistance increases, see figure 7.2a.

In figure 7.2b, the error introduced by neglecting the ionic contribution is shown together with the difference of R_3 and R_1. At low pH_2O with high LSR_{ct} the difference (R_3-R_1) is positive, indicating charge transfer limitation; at high pH_2O with low LSR_{ct} the difference is negative, indicating transport limitation. Thus, for transport limitation conditions, the error made by neglecting the contribution of ionic transport is largest due to (i) a high impedance loss in the electrolyte and (ii) a small penetration depth leading to an overestimation of l_{TPB} in equation 7.11.

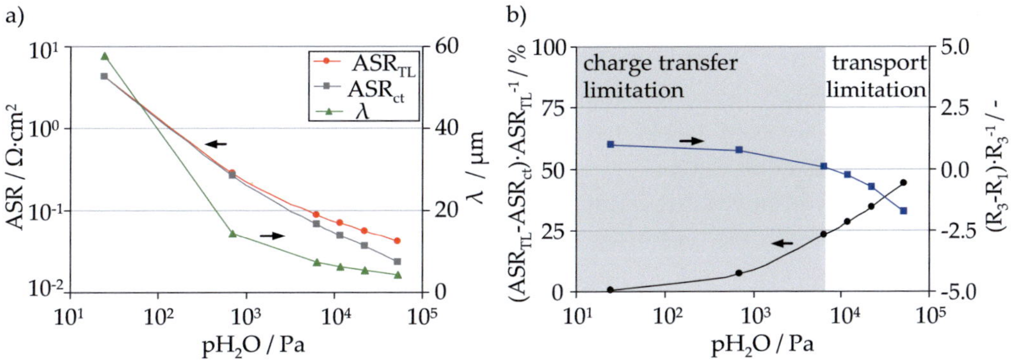

figure 7.2: a) Comparison of ASR values calculated with contribution of ionic transport ASR_{TL} (equations 7.8 to 7.10) and neglecting ionic transport ASR_{ct} (equation 7.11). Additionally, the penetration depth is indicated. LSR_{ct} values are taken from figure 5.1 at $pH_2 = 9.0 \cdot 10^3$ Pa, $T = 800$ °C; $\sigma_{ion,eff} = 1.11\ S \cdot m^{-1}$ and $l_{TPB} = 2.53 \cdot 10^3 m \cdot cm^{-2}$, and b) deviation between values of ASR_{TL} and ASR_{ct} from a) together with the subtraction of R_3 and R_1, indicating charge transfer limitation for $pH_2O < 7.0 \cdot 10^3$ Pa and transport limitation for higher pH_2O.

Neglecting the ionic transport contribution can lead to an error of almost 50 % in ASR determination, see figure 7.2b. Hence, for a meaningful comparison of the electrochemical performance of patterned and cermet anodes, the calculation of the transmission line resistance R_{TL} including charge transfer and ionic transport is essential.

7.3 Apparent Parameter Dependencies

The measurement setup and most importantly the analysis of the impendance spectra of Ni/8YSZ cermet anodes have an influence on the observed parameter dependencies: they may either describe the combined dependencies of charge transfer and ionic transport or the individual dependencies. Sonn et al. have used a transmission line model for the analysis of the electrochemical characterization of symmetrical Ni/8YSZ cermet anodes [35]. This enabled the separation of the contributions of charge transfer and ionic transport. Additionally, the activation energies of both contributions have been determined individually. However, in most cases this separation is not performed, but the reported parameter dependencies include contributions of charge transfer and ionic transport. These parameter dependencies are then referred to as apparent parameter dependencies. Among others, this applies to the data reported by Leonide for the electrochemical characterization of the anode function layer of anode supported cells [33, 53, 99]

The difference between apparent dependency and pure dependency of charge transfer on gas composition and on temperature is a function of the limitation regime and hence the R_3/R_1 ratio.

Gas Composition Dependency

The ionic conductivity in the electrolyte is invariant to changes in the gas composition (pH_2-pH_2O or pCO-pCO_2), whereas charge transfer is strongly affected, see chapters 5 and 6. Hence, the apparent dependency will be reduced compared to the parameter dependency of pure charge transfer. In the special case of transport limitation ($R_1 >> R_3$), where the attenuation of the effect of changes in the gas composition is strongest, the apparent parameter dependencies amount only for half of the parameter dependencies of pure charge transfer, see equations 7.12 and 7.13.

Thus, depending on the R_3/R_1 ratio in technical Ni/8YSZ cermet anodes, the apparent parameter dependency a_{app} is expected to lie within an interval of $0.5 \cdot a \leq a_{app} \leq a$ (same applies for b, c, and d).

$$Z_{TL}\left(pH_2, pH_2O\right) = \sqrt{R_1 \cdot R_3} = \sqrt{R_1 \cdot c_7} \cdot pH_2^{-\frac{a}{2}} \cdot pH_2O^{-\frac{b}{2}} \qquad 7.12$$

$$Z_{TL}\left(pCO, pCO_2\right) = \sqrt{R_1 \cdot R_3} = \sqrt{R_1 \cdot c_8} \cdot pCO^{-\frac{c}{2}} \cdot pCO_2^{-\frac{d}{2}} \qquad 7.13$$

Activation Energy

The temperature dependency of the ionic transport resistance is weaker than for the charge transfer resistance. Hence, the apparent activation energy will be reduced compared to the activation energy of pure charge transfer. In the special case of transport limitation, the apparent activation energy $E_{act,app} = E_{act,RTL}$ is calculated as the mean value of the activation energies of ionic transport $E_{act,R1}$ and charge transfer $E_{act,R3}$, see equation 7.14. In the case of charge transfer limitation, the apparent activation energy lies between the two activation energies $E_{act,R1}$ and $E_{act,R3}$.

$$E_{act,R_{TL}} = \frac{E_{act,R_1} + E_{act,R_3}}{2} \qquad 7.14$$

As a consequence of the different activation energies, a change in the penetration depth λ with temperature is expected and was confirmed experimentally by Sonn et al. [35].

7.4 Comparison of Electrochemical Performance

As shown in equations 7.8 - 7.10, the calculation of the transmission line resistance at DC requires the knowledge of LSR_{ct} together with $l_{TPB,v}$ and $\sigma_{ion,eff}$. LSR_{ct} values have been determined in this work for parameter variations of T in H_2-H_2O and CO-CO_2 atmosphere as well as of the partial pressures pH_2, pH_2O, pCO and pCO_2. The parameters $l_{TPB,v}$ and $\sigma_{ion,eff}$ are unknown and therefore estimated based on modeling and experimental studies respectively.

The value of $l_{TPB,v}$ can be estimated using 3D reconstruction [103-106] or using microstructure models [39]. Besides the material composition that is decisive for the amount of TPB [1, 39], the percolation percentage is an important factor in order to consider only the active TPB. Results of different studies are summarized in section 9.3. With respect to the differences in material composition, they show good agreement with $l_{TPB,v}$ values between 10^{12} - 10^{13} m^{-2}. In this work, $l_{TPB,v}$ was estimated based on a simulation of the electrode with COMSOL Multiphysics according to the simulation studies reported by Rüger et al. [107, 108]. For a particle size of 500 nm and a material composition of 48 - 30 - 22 vol% YSZ - Ni - pore, which corresponds to the anode functional layer (AFL) characterized in the studies of Leonide et al. [33, 98], a value of $l_{TPB,v} = 3.61 \cdot 10^{12}$ m^{-2} was obtained. For a material composition of screen printed Ni/8YSZ cermet anodes with a material composition of 35 - 30 - 35 vol % YSZ - Ni - pore [35], a slightly higher value of $4.21 \cdot 10^{12}$ m^{-2} was obtained.

The ionic conductivity of the YSZ skeleton in a cermet anode which is denoted effective ionic conductivity $\sigma_{ion,eff}$, is different from the bulk conductivity $\sigma_{ion,bulk}$. This is taken into account by the structural factor f as a function of porosity ε and tortuosity τ of the YSZ skeleton [102]:

$$\sigma_{ion,eff} = f \cdot \sigma_{ion,bulk} = \frac{1-\varepsilon}{\tau} \sigma_{ion,bulk} \tag{7.15}$$

Furthermore, the ionic conductivity exhibits a strong degradation due to a phase transition from cubic to tetragonal leading to Y-depletion and Y-enriched regions with lower ionic conductivity [109-111]. In Ni/8YSZ cermet anodes the phase transition and hence the degradation is assumed to be induced by the diffusion of NiO into the YSZ during sintering and subsequent reduction of NiO to Ni in reducing atmosphere.

In this work, $\sigma_{ion,eff}$ values were taken from conductivity measurements of an anode substrate, where Ni has been removed after sintering. At 800 °C, a value of $\sigma_{ion,bulk}(800\ °C) = 4.61\ S \cdot m^{-1}$ has been obtained, yielding a value of $\sigma_{ion,eff}(800\ °C) = 1.106\ S \cdot m^{-1}$ with $\varepsilon = 0.52$ and $\tau = 2$. After thermal exposure at 950 °C for 100 h in reducing atmosphere, the ionic conductivity has degraded considerably to $\sigma_{ion,bulk,d}(800\ °C) = 1.49\ S \cdot m^{-1}$ and thus $\sigma_{ion,eff,d} = 0.357\ S \cdot m^{-1}$. The ASC cells characterized by Leonide et al. have not been exposed to temperatures exceeding 870 °C [90], the degradation of the ionic conductivity is therefore expected to be less pronounced.

The results of calculation of R_{TL} obtained using the LSR_{ct} values combined with $l_{TPB,v}$ and $\sigma_{ion,eff}$ are shown in figure 7.3 to figure 7.6 compared to experimental data from anode supported cells characterized in H_2-H_2O and CO-CO_2 atmosphere reported in [53, 90]. These samples are composed of a 7 µm thick anode functional layer on the anode substrate, hence L = 7 µm.

In figure 7.3 the dependencies of the resistances on pH_2O and pH_2 are given. The closed symbols refer to the non-degraded ionic conductivity, the open symbols to the degraded value after thermal exposure at 950 °C for 100 h. The error bars indicate the effect of an over- or underestimation of the $l_{TPB,v}$ of 50 %. As expected, the dependencies of the transmission line resistance decrease compared to the dependencies of the LSR_{ct}, due to the contribution of gas composition independent ionic transport in the YSZ skeleton. For non-degraded ionic conductivity, the resistance is slightly higher than the value of the ASR_{AFL} obtained for cermet anodes. The deviation increases when the degraded ionic conductivity is used. However, even in the case of degraded ionic conductivity, the deviation of the resistances is below one order of magnitude.

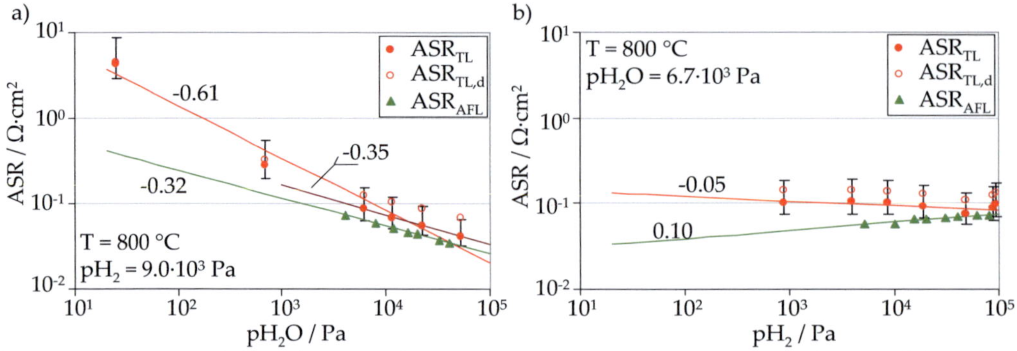

figure 7.3: Comparison of electrochemical performance of patterned and cermet anodes as a function of a) pH_2O at constant pH_2, and b) pH_2 at constant pH_2O. The subscript 'TL' refers to results from transmission line model calculation, 'TL,d' to a transmission line model including the degraded ionic conductivity, and 'AFL' to results reported for an anode functional layer by Leonide [33].

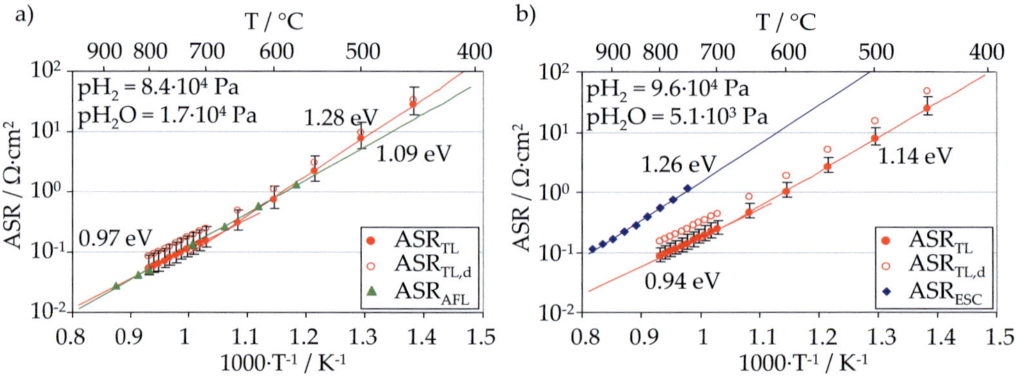

figure 7.4: Comparison of electrochemical performance of patterned and cermet anodes for H_2-H_2O atmosphere as a function of temperature: a) comparison to data reported by Leonide for the anode functional layer of an anode supported cell [90], calculation with L = 7 μm, and b) comparison to data reported by Sonn for a screen printed cermet anode [35] with L = 25 μm. The subscript 'TL' refers to results from transmission line model calculation, 'TL,d' to a transmission line model including the degraded ionic conductivity, 'AFL' to results reported for an anode functional layer by Leonide [90], and 'ESC' to results reported for a screen printed cermet anode by Sonn [35].

For the activation energy in H_2-H_2O atmosphere, the comparison to data of a screen printed Ni/8YSZ cermet anode characterized by Sonn et al. [35] is possible in addition to the comparison of electrochemical performance to values of the AFL, see figure 7.4. For the gas composition of figure 7.4a activation energies of 0.97 eV for 700 - 800 °C and 1.28 eV for 450 - 700 °C were obtained. The value of 1.09 eV for the AFL lies within these values. Furthermore, the absolute ASR values are in very good agreement. The ASR values of the screen printed anode are higher than the values calculated based on LSR_{ct}. However, the deviation is still less than one order of magnitude, see figure 7.4b. Also the activation energy of 1.26 eV is higher than 0.94 and 1.14 eV obtained with the transmission line calculation. Note that the activation energy obtained with the transmission line model changes with gas composition due to the different R_3/R_1 ratio.

The comparison of electrochemical performance in CO-CO_2 atmosphere shows the same trend as for H_2-H_2O with good agreement of the absolute ASR values, see figure 7.5. The parameter dependencies of ASR_{TL} are attenuated compared to those of LSR_{ct}, but still stronger than for the experimentally determined values for the AFL. For the pCO variation, good agreement is observed for pCO > 10^4 Pa with c = 0.06 for ASR_{AFL} and c = 0.07 obtained for ASR_{TL}, see figure 7.5b.

The behavior for temperature dependency in CO-CO_2 atmosphere is similar, see figure 7.6. The ASR_{AFL} values are even slightly higher than values of ASR_{TL}. The activation energies are in good agreement with 1.29 eV for the transmission line calculation of the patterned anode, and 1.23 eV for the cermet anode.

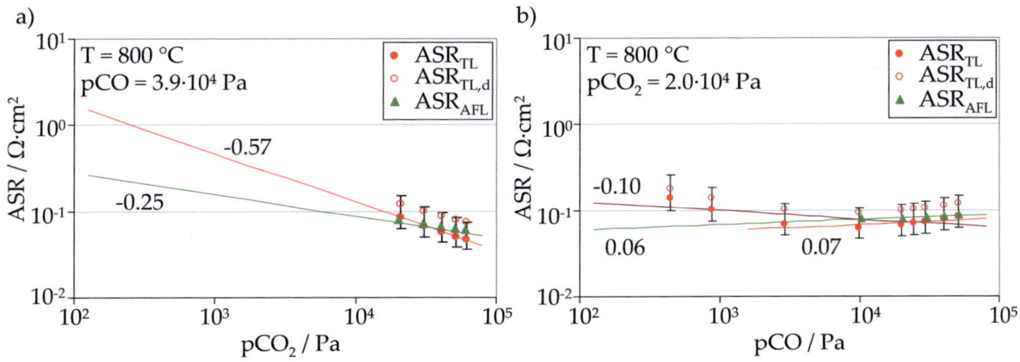

figure 7.5: Comparison of electrochemical performance of patterned and cermet anodes as a function of a) pCO_2 at constant pCO, and b) pCO at constant pCO_2. The subscript 'TL' refers to results from transmission line model calculation, 'TL,d' to a transmission line model including the degraded ionic conductivity, and 'AFL' to results reported for an anode functional layer by Leonide [98].

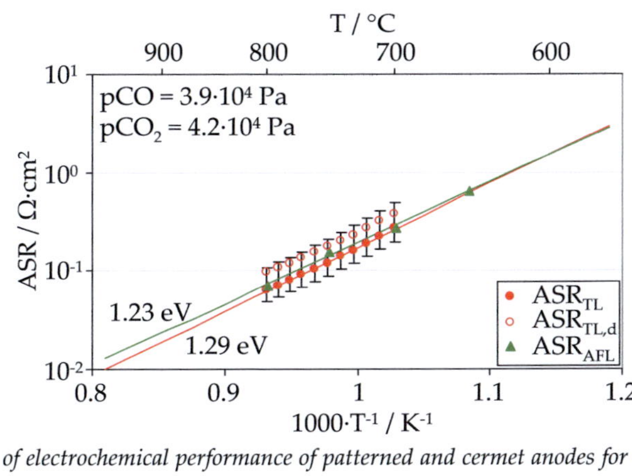

figure 7.6: Comparison of electrochemical performance of patterned and cermet anodes for CO-CO_2 atmosphere as a function of temperature. The subscript 'TL' refers to results from transmission line model calculation, 'TL,d' to a transmission line model including the degraded ionic conductivity, and 'AFL' to results reported for an anode functional layer by Leonide [98].

A summary of the dependencies of LSR_{ct}, ASR_{TL}, $ASR_{TL,d}$ and ASR_{AFL} on the parameters T, pH_2, pH_2O, pCO, and pCO_2 is given in table 7.3. It is clearly observable that the parameter dependencies of ASR_{TL} have decreased compared to those of LSR_{ct}, since ionic transport in the electrolyte is invariant to gas composition and only slightly temperature activated. In the case of a calculation with the degraded ionic conductivtity yielding $ASR_{TL,d}$, the dependencies decrease further.

table 7.3: Comparison of parameter dependencies of LSR_{ct} and the resulting R_{TL} to the apparent parameter dependencies reported for a Ni/8YSZ cermet anode (ASR_{AFL}) by Leonide [90, 98]. a and b refer to pH_2 and pH_2O dependencies following equations 5.3 and 5.4, c and d refer to pCO and pCO_2 dependencies following equations 6.2 and 6.3.

Parameter dependency	Range	LSR_{ct}	R_{TL}	$R_{TL,d}$	ASR_{AFL} [33, 98, 99]
b [-]	total:	0.68 ± 0.04	0.61 ± 0.05	0.54 ± 0.06	0.33 ± 0.008
	$>10^3$ Pa:		0.35 ± 0.004	0.28 ± 0.003	
a [-]		0.07 ± 0.04	0.05 ± 0.03	0.04 ± 0.02	- 0.10 ± 0.01
E_{act} (H_2-H_2O) [eV]	700 - 800°C:	1.01 ± 0.01	0.97 ± 0.008	0.97 ± 0.006	1.09 ± 0.02
	450 - 700°C:	1.37 ± 0.02	1.28 ± 0.02	1.21 ± 0.02	
d [-]	$pCO=3.9\cdot10^4$ Pa:	0.79 ± 0.02	0.57 ± 0.02	0.44 ± 0.02	0.25 ± 0.006
	$pCO=1.0\cdot10^4$ Pa:	0.61 ± 0.03	0.51 ± 0.05	0.42 ± 0.04	
c [-]	$pCO_2=5.1\cdot10^4$ Pa:	0.24 ± 0.05	0.20 ± 0.03	0.15 ± 0.03	- 0.059 ± 0.016
	$pCO_2=2.0\cdot10^4$ Pa:	0.11 ± 0.05	0.1 ± 0.04	0.08 ± 0.03	
E_{act} (CO-CO_2) [eV]	700 - 800°C:	1.42 ± 0.02	1.29 ± 0.02	1.22 ± 0.01	1.23 ± 0.07

7.5 Discussion

The basic question in the interpretation of experimental data of patterned Ni anodes is whether findings on the elementary kinetic reaction mechanism can be transferred to technical Ni/8YSZ cermet anodes. Among others, this involves the possible influence of impurity barriers for the charge transfer reaction at the TPB. Due to the different surface to volume ratio between patterned and cermet anodes, the effect of impurities on the electrochemistry of patterned anodes is considered stronger than for the nano- or microstructured cermet anodes, see section 9.2. The comparison of the electrochemical performance of patterned and cermet anodes gives evidence to this question.

The comparison performed in this work yields a good agreement of the absolute ASR values as well as of most parameter dependencies to experimental data of a Ni/8YSZ cermet anode. Yet, for the pCO_2 variation a stronger dependency was obtained for the data of patterned anodes than for the cermet anode.

Additionally, the calculation according to the transmission line model depends on several parameters that need to be estimated ($l_{TPB,v}$, $\sigma_{ion,bulk}$, and $\sigma_{ion,eff}$). Hence, the result is subject to uncertainties that will be discussed in the following sections.

Uncertainties in $l_{TPB,v}$

Two methods have been presented for the determination of $l_{TPB,v}$ including 3D reconstruction and microstructural models that both show disadvantages. The result from 3D reconstruction is only valid for the analyzed material composition and requires the reconstruction of the actually characterized AFL layer, which has not been performed so far. Furthermore, image analysis is strongly dependent on the contrast between the three different phases and the consecutive segmentation step. The size of the analyzed image influences the reproducibility, and therefore a representative volume element needs to be determined [106]. Additionally, the percolation percentage is difficult to determine. In microstructural models the particles are typically represented by spheres or cubes and geometry changes due to sintering (e.g. agglomeration) are only included to a limited extent. Therefore, the obtained values of $l_{TPB,v}$ typically represent an upper limit. The value of $l_{TPB,v} = 3.61 \cdot 10^{12}$ m^{-2}, which has been chosen in this work based on a microstructural model, appears reasonable with respect to literature data of several studies that all show similar values in the order of 10^{12} m^{-2} for different material compositions, see section 9.3.

Uncertainties in σ_{ion}

For the ionic conductivity, the uncertainties are related to the experimental identification of $\sigma_{ion,bulk}$, its degradation and the calculation of $\sigma_{ion,eff}$ using the structural factor. For the calculation of the effective ionic conductivity $\sigma_{ion,eff}$, all parameters – the structural factor f, the porosity ε, and the tortuosity τ – are strongly dependent on the material composition. For the AFL, values of $\varepsilon = 0.52$ and $\tau = 2$ ($f = 0.24$) have been used in accordance with a study representing τ as a function of ε [107]. For the comparison to the screen printed Ni/8YSZ cermet anode, values of $\varepsilon = 0.65$ and $\tau = 2.5$ ($f = 0.14$) were used. The degradation of the ionic conductivity was shown to be caused by a decomposition of the YSZ matrix from cubic to tetragonal with Y-enriched and Y-depleted regions that both exhibit a reduced ionic conductivity compared to the original bulk composition [109-111]. This decomposition is accelerated by the diffusion of NiO into the YSZ matrix and subsequent reduction to Ni in reducing atmosphere. Therefore, the degradation is especially relevant in Ni/8YSZ cermet anodes and strongly dependent on sintering temperature as well as temperature and atmosphere during electrochemical characterization, with stronger degradation in reducing atmosphere [111]. In this work, two different values for the ionic conductivity were applied: (i) the non-degraded ionic conductivity $\sigma_{ion,bulk}$ that has been experimentally determined for temperatures of 600 – 950 °C in reducing atmosphere, and (ii) the degraded ionic conductivity $\sigma_{ion,bulk,d}$ that has been experimentally determined for the same temperature range after thermal exposure at 950 °C for 100 h in pure H_2.

Absolute ASR Values

The comparison of ASR_{TL} and ASR_{AFL} values shows good agreement, see figure 7.4a. Better agreement is obtained with values for non-degraded $\sigma_{ion,eff}$. This is attributed to the reduced temperature during characterization with a maximum temperature of 870 °C in the study of Leonide, whereas the degraded ionic conductivity refers to thermal exposure at 950 °C and consequently a stronger degradation. However, even non-degraded ASR_{TL} values are slightly above ASR_{AFL}, especially in H_2-H_2O atmosphere. An explanation involves an under-estimation of $l_{TPB,v}$, which is rather improbable (see discussion above). Instead, the opposite case of a blocking of parts of the TPB seems more likely, leading to a further increase in ASR_{TL}. Hence, with the actual uncertainties in $l_{TPB,v}$ and $\sigma_{ion,eff}$, it cannot be excluded that a slight difference between patterned and cermet anode reaction mechanisms persists e.g. through a stronger effect of impurities at the TPB of patterned anodes, see section 3.4.

The deviation between ASR_{TL} and data obtained for screen printed cermet anodes reported by Sonn (ASR_{ESC}) is higher, though still within one order of magnitude, see figure 7.4b. For this data, the separation of ionic transport contribution R_1 and charge transfer contribution R_3 has been performed, hence a comparison of the individual values can be performed, see figure 7.7. The $R_{1,ESC}$ values show good agreement with $R_{1,d,PA}$ for degraded ionic conductivity, whereas the calculation with the non-degraded ionic conductivity results in too low values, see figure 7.7a. This is expected, as the screen printed electrode has been characterized at 950 °C and the ionic conductivity is therefore strongly affected by degradation.

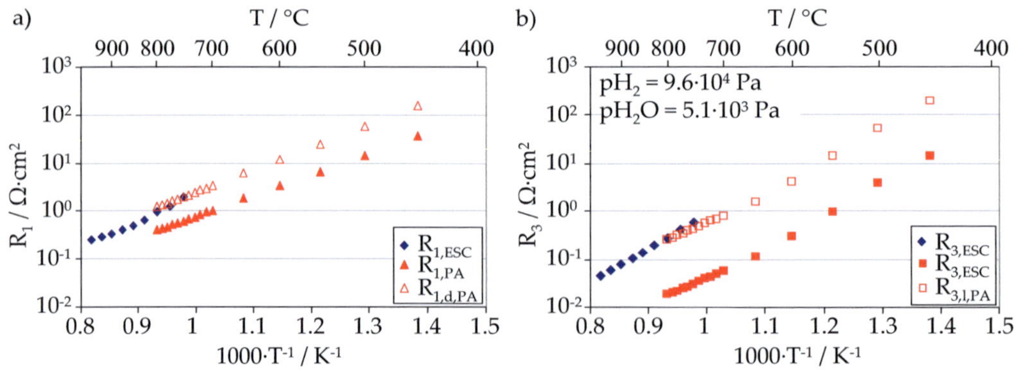

figure 7.7: Comparison of values of R_1 and R_3 calculated in this work with values reported by Sonn [35] for a screen printed Ni/8YSZ cermet anode as a function of temperature: a) good agreement for values of R_1 is obtained when using the degraded ionic conductivity, yielding $R_{1,d,PA}$. Using the standard ionic conductivity, the obtained $R_{1,PA}$ is inferior to $R_{1,ESC}$ (factor 3 to 4), and b) a strong deviation in R_3 is observed between data by Sonn and the calculation in this work when using $l_{TPB,v} = 4.2 \cdot 10^{12}$ m^{-2} ($R_{3,PA}$). With a correction of $l_{TPB,v}$ to $3.0 \cdot 10^{11}$ m^{-2}, good agreement is obtained. The subscript 'PA' refers to data from this work with patterned anode and 'ESC' to the cermet anode data reported by Sonn with L = 25 μm.

The R_3 values strongly deviate by one order of magnitude, where data by Sonn is higher than $R_{3,PA}$ calculated using LSR_{ct} according to the equations in table 7.1. Using $l_{TPB,v} = 3.0 \cdot 10^{11}$ m^{-2} instead of $4.2 \cdot 10^{12}$ m^{-2} leads to a good agreement (see figure 7.7b, open symbols). However, this would imply that only 7 % of the entire TPB actively contributes to the electrochemical reaction, which is an improbably low value. Another explanation may be that due to the different temperature regimes of characterization (450 °C ≤ T ≤ 800 °C for patterned anodes and 750 °C ≤ T ≤ 950 °C for the cermet anode), different reaction mechanisms prevail. For example, the kink observed in the activation energy of the patterned anodes (figure 5.3) was attributed to a change in the reaction mechanism with a limitation by two hydrogen spillover mechanisms at T > 700 °C [13, 112]. A similar observation was made in the impedance modeling study by Gewies et al. [39]: the impedance data reported Sonn was only reproducible by changing the reaction mechanism compared to the model adapted for data by Bieberle for patterned Ni anodes [6].

Parameter Dependencies of ASR

In the case of temperature dependency, good agreement is found for the activation energies. The activation energy of ASR_{AFL} in H_2-H_2O atmosphere is between the two values determined for patterned anodes for temperatures above and below 700 °C. In CO-CO_2 atmosphere, good agreement is observed as well, see table 7.3.

In the case of gas dependencies, deviations persist for pH_2O and pCO_2, with ASR_{TL} dependencies by a factor of 1.5 to 3.4 higher than the experimentally observed dependencies. Using the degraded ionic conductivity (thus higher R_1), the dependencies decrease and approach the values for the cermet anode, see table 7.3. This is an indication that the ratio of R_1/R_3 for the experimentally characterized Ni/8YSZ cermet anodes is higher than in the transmission line calculation. This means that in this work either R_3 is overestimated or R_1 is underestimated, or both. On the one hand, this may be attributed to a false estimation of the calculation parameters $l_{TPB,v}$, ε and τ, as well as the degradation of the ionic conductivity. On the other hand, this can be explained by a higher LSR_{ct} at the TPB of patterned anodes than of cermet anodes, indicating a blocking by impurities.

The dependencies of pH_2 and pCO are weaker and in general exhibit higher standard deviations. Therefore, a detailed comparison to cermet anode data is not possible.

Good agreement has been observed between the performance and parameter dependencies of patterned and cermet anodes. However, slightly different reaction mechanisms cannot be excluded due to (i) the tendency to higher ASR values for patterned anodes and (ii) the observation of weaker parameter dependencies than expected from transmission line calculation. A possible difference may include an additional loss mechanism in the case of patterned Ni anodes due to impurity barriers at the TPB.

7.6 Conclusions

The comparison of the electrochemical performance of patterned Ni anodes and Ni/8YSZ cermet anodes was enabled by the use of a transmission line model. Therein, ASR_{TL} values have been calculated considering charge transfer at the TPB and oxygen ion transport in the YSZ. The comparison was applied to data reported for the anode functional layer of an anode supported cell by Leonide and the screen printed cermet anode of an electrolyte supported cell by Sonn.

Very good agreement was obtained between the theoretical values and experimentally determined data by Leonide. This holds especially for absolute ASR values and the activation energy in H_2-H_2O and CO-CO_2 atmosphere. For the dependencies on gas composition, the transmission line calculation results in stronger dependencies than experimentally observed. This is attributed to a too high R_3/R_1 ratio that may be explained by (i) a false estimate of the calculation parameters ($l_{TPB,v}$, $\sigma_{ion,bulk,d}$, ε, and τ) or (ii) a slight difference in the reaction mechanism.

The stronger deviations to data by Sonn – that are still within one order of magnitude – are attributed to the different temperature during electrochemical characterization and may include (i) an accelerated degradation of the ionic conductivity or (ii) a difference in reaction mechanism as suggested by modeling studies.

Nevertheless, the good agreement with data by Leonide for the same temperature of characterization (T = 800 °C) indicates a similar reaction mechanism for patterned anodes and cermet anodes, even though an additional limitation due to impurities at the TPB of patterned anodes cannot be excluded with respect to the uncertainties in the estimation of the transmission line model parameters $l_{TPB,v}$ and $\sigma_{ion,eff}$. Hence, the applicability of patterned anodes for the investigation of the electrochemical oxidation of the fuel gas at SOFC anodes is affirmed.

8 Discussion and Summary

8.1 Concluding Discussion

In this section, the findings of the individual chapters of this work are discussed concludingly with respect to the reaction mechanism of the electrochemical oxidation of H_2 and CO at SOFC anodes. In this work, a deeper understanding is gained by the combination of results on different aspects of patterned anode characteristics as well as of results obtained with different anode concepts. These combinations include:

(i) The comparison of H_2-H_2O results to different patterned Ni anode studies in literature.

(ii) The comparison of CO-CO_2 results to data for the Ni grid electrode reported by Boulenouar.

(iii) The combined comparison of electrochemical performance and impurity features to results reported for Ni point anodes.

(iv) The comparison of results for H_2-H_2O and CO-CO_2 atmosphere.

(v) The comparison to results of technical Ni/8YSZ cermet anodes using a transmission line model calculation.

The large deviations found between the studies with patterned Ni anodes [6, 7, 9, 34] – two orders of magnitude for LSR and considerable scatter for the parameter dependencies pH_2, pH_2O and T – indicate a significant difference in the characterization of the samples even though seemingly similar experimental conditions prevailed, see section 2.1. In this work, a lot of effort was directed to assure stability of the patterned anode and thus reproducibility of measurement results:

(i) Different facrication procedures have been investigated, enabling the reproducible preparation of stable patterned anodes with good edge quality [77].

(ii) A detailed stability analysis of Ni thin films as a function of layer thickness, temperature, gas composition and exposure time was performed, yielding an ideal Ni layer thickness of 800 nm for the electrochemical characterization up to temperatures of 800 °C and varying water vapor content [113].

(iii) The reproducibility of measurement results was studied and revealed an initial activation of anode performace, a water vapor dependent degradation rate of LSR_{ct} as well as slow relaxation processes with settling times up to 5 h after a change in gas composition and up to 20 h after the recording of a current-voltage curve [58]. The reproducible recording of parameter variations was enabled by a standard initial heat treatment before the actual electrochemical characterization and long waiting times after each gas variation step and polarization.

(iv) The impurity features of the patterned Ni anode were studied in a detail impurity analysis including AFM, XPS and TOF-SIMS and demonstrated the low impurity level of the patterned Ni anodes characterized in this work [87].

Hence, this work differs significantly from patterned Ni anode studies in literature that did not specify most of these points. Especially the initial heat treatment, the procedure of electrochemical characterization as well as the impurity level are decisive and may explain the observed considerable deviations in experimental results.

Furthermore, for the first time the pure charge transfer contribution LSR_{ct} of patterned Ni anodes and its parameter dependencies have been determined. This was enabled by the separation of the three loss mechanisms – charge transfer at the patterned Ni anode, gas diffusion and charge transfer in the counter electrode – based on the analysis of impedance spectra with DRT calculation and CNLS fits.

In the case of CO-CO_2 atmosphere, sparse literature exists on parameter dependencies with model anodes. The observed strong deviation between results from this work and data from Ni grid electrodes by Boulenouar [47], that exceed one order of magnitude, are attributed to the significant differences between the two electrode concepts: in order to assure contacting of Ni grid and YSZ surface, Boulenouar applied a sintering step at 1400 °C. However, it is assumed that during this heat treatment a large amount of impurities from electrolyte and electrode bulk segregates at the TPB. Note that the purity of both materials is not specified.

A very good agreement to results of point anodes was obtained with a deviation of LSR within one order of magnitude, even though differences in the anode concept persist, e.g. a strong difference in l_{TPB} and a different method of contacting. This emphasizes the similar and low impurity level that was achieved for both sample types by careful choice of materials and measurement setup. Furthermore, it affirms that the description and consideration of the dynamic processes during the recording of impedance spectra yields reproducible experimental data. By contrast, no detailed description of dynamic processes is given for most literature studies on patterned Ni anodes.

The comparison of results for H_2-H_2O and CO-CO_2 atmosphere is of special interest in the investigation of possible reaction mechanisms. On the one hand, the observed unexpected similarities of absolute LSR_{ct} values and parameter dependencies on partial pressures of reactant and reaction product as well as temperature indicate similar reaction mechanisms in both atmospheres. On the other hand, differences exist for the degradation and relaxation behavior. In H_2-H_2O atmosphere, a higher degradation rate is obtained which is attributed to a removal of Ni by volatile $Ni(OH)_2$ and / or diffusion of impurities to the TPB. In any case, both effects are very small on the scale of the Ni structure, since SEM analyses of patterned Ni anodes after electrochemical characterization show no discernable difference for samples exposed to the two atmospheres [114]. Additionally, in H_2-H_2O atmosphere slow relaxation processes with long settling times are observed for changes in the gas composition (especially for pH_2), while stabilization is fast in CO-CO_2 atmosphere with settling times of few minutes. The long settling times in H_2-H_2O are most likely attributed to changes in bulk species concentration and may be attributed to a limitation by an interstitial charge transfer. In the case of CO-CO_2 atmosphere, this mechanism is not expected due to the large molecule size, and may therefore explain the differences in observed relaxation behavior.

So far, modeling studies of the CO oxidation were mostly limited to few reaction mechanisms [14, 61]. With the similarity in performance between the two atmospheres, special focus will need to be directed to the comparison of CO spillover and H spillover mechanisms, since the latter is in favor of most modeling studies for the H_2 oxidation [13, 16, 54] and CO spillover corresponds to H spillover in CO-CO_2 atmosphere. Additionally, uncertainty persists with respect to a possible interstitial charge transfer. On the one hand, interstitial charge transfer is considered ineffective in CO-CO_2 atmosphere due to the large molecule size. On the other hand, interstitial charge transfer is advocated for point anodes in H_2-H_2O atmosphere [71]. Based on the initial behavior of decreasing LSR and the dependency of LSR on electrolyte purity, the reaction is interpreted as proceeding through the electrode-electrolyte interface with additional loss mechanisms due to impurities at the interface. Furthermore, interstitial charge transfer is suggested to proceed in parallel to a surface charge transfer mechanisms [59]. In a modeling study of the interstitial charge transfer, Vogler showed that the mechanism is in principle feasible, but probably restricted to small current densities [57]. However, the underlying diffusion coefficients are subject to large uncertainties of up to 3 orders of magnitude [16].

Therefore, in H_2-H_2O atmosphere an interstitial charge transfer mechanisms may account for parts of the current density in parallel to a surface spillover mechanisms. It would then be responsible for the observed slow relaxation mechanisms observed in H_2-H_2O atmosphere, in contrast to CO-CO_2 where interstitial charge transfer is not relevant.

Finally, the comparison to cermet anode performance completes the investigation with an evaluation of applicability and significance of the analysis of the anode reaction mechanisms by means of patterned Ni anodes. For this comparison, a transmission line model calculation has been applied that allows to combine the contributions of charge transfer and ionic transport in the YSZ matrix. Results show very good agreement with data reported for the anode functional layer of an anode supported cell by Leonide both for H_2-H_2O and CO-CO_2 atmosphere. Especially, good agreement is obtained for absolute ASR values and for the activation energy, whereas the dependencies on the partial pressures of H_2, H_2O, CO and CO_2 are more pronounced for the calculated ASR_{TL} values than for the experimentally determined ASR_{AFL}. These parameter dependencies are strongly dependent on the R_3/R_1 ratio, with higher dependencies for a higher ratio. This is due to the gas composition invariant total ionic transport resistance R_1, whereas the total resistance of charge transfer R_3 is strongly dependent on gas composition. Hence two explanations for the stronger theoretical parameter dependencies exist:

(i) R_3 may be overestimated in the transmission line calculation either by an erroneous estimate of the TPB length or impurities at the TPB that lead to an additional loss mechanism. This assumption is based on the finding that different impurity situations exist for compact electrolytes in the case of patterned anodes compared to the porous anode structure of cermet anodes: While avoidance of a complete coverage of the electrolyte surface with impurities seems almost impossible in the case of model anodes, technical cermet anodes are less prone to surface coverage with impurities due to the three orders of magnitude higher surface to volume ratio (see also the calculcation in section 9.2).

(ii) R_1 may be underestimated by an insufficient consideration of the degradation in ionic conductivity or due to an erroneous calculation of the effective ionic conductivity using rough estimates of the cermet parameters porosity ε and tortuosity τ.

Additionally, the YSZ electrolyte at the TPB of patterned and cermet anodes may be different due to the fabrication procedure: whereas patterned Ni anodes are prepared by direct application of the Ni thin film on the YSZ electrolyte, technical cermet anodes are prepared based on NiO with subsequent sintering and reduction in order to realize a porous Ni/8YSZ structure. During sintering, NiO diffuses into the YSZ volume and – upon reduction to Ni – accumulates at the grain boundaries, causing an accelerated degradation of ionic conductivity [109-111]. This difference at the TPB may result in an increased LSR and thus R_3 for cermet anodes. However, this is not affirmed by the calculation in this work.

The results with similar parameter dependencies and the good agreement of absolute ASR values indicate that the same reaction mechanism prevails for patterned anodes and for technical cermet anodes. However, slight differences e.g. an additional loss mechanism due to an impurity layer at the TPB of patterned anodes cannot be excluded with respect to the uncertainties for the parameters $l_{TPB,v}$ and $\sigma_{ion,eff}$ in the transmission line calculation. It can be concluded that – provided that sample stability and measurement reproducibility are ensured – patterned anodes represent a meaningful method for the analysis of the electrochemical oxidation of the fuel gas at SOFC anodes.

8.2 Summary

In this work, valuable insight into the reaction mechanism of the electrochemical oxidation of H_2 and CO at SOFC anodes was gained by (i) the establishment of comprehensive data sets of the electrochemical characterization of patterned Ni anodes in H_2-H_2O and CO-CO_2 atmosphere and (ii) the extensive comparison of results to a variety of different anode aspects and anode concepts.

In the introductory literature review, the need of a new data set with large range of parameter variation of pH_2, pH_2O, T and η was demonstrated by the significant deviations in literature existing for the electrochemical characterization of patterned Ni anodes in H_2-H_2O atmosphere. Additionally, explanations for these differences are essential. Furthermore, the electrochemical characterization needs to be extended to CO-CO_2 atmosphere in order to study the reaction mechanism of reformate gases.

The reproducible fabrication of stable patterned Ni anodes with sharp edges was realized by a sequence of Ni sputtering on a structured photoresist followed by lift-off. A thickness of 800 nm has been shown to ensure the stability of the Ni structure during electrochemical characterization up to temperatures of 800 °C in reducing atmosphere.

The electrochemical characterization was performed in a gas tight measurement setup that allows a precise adjustment and monitoring of the gas composition. Large variations of the water vapor content of 0.02 % $\leq \beta_{rel} \leq$ 88 % were realized by an upstream combustion chamber. Typical errors induced by reference electrodes were avoided by the characterization against a Ni/8YSZ cermet electrode with known contribution.

In data analysis, the determination of the 'pure' LSR_{ct} value of patterned Ni anodes was enabled for the first time by the combined evaluation of impedance spectra with DRT calculation and CNLS fits, yielding three separate processes attributed to (i) charge transfer reaction at the patterned anode ($R_{A\beta}$), (ii) gas diffusion ($R_{A\alpha}$), and (ii) charge transfer at the counter electrode ($R_{A\gamma}$).

Even though impurities cannot be avoided entirely for Ni electrodes at elevated temperatures of 800 °C, in a detailed impurity analysis their amount has been shown to be very low. This was realized by the careful choice of clean materials for the sample and the measurement setup.

The detailed analysis of measurement reproducibility revealed that nominally stable patterned Ni anodes are undergoing considerable relaxation and degradation processes with time constants in the order of several hours:

(i) An initial decrease of LSR_{ct} by 60 % to an average value of 54 Ω·m with stabilization within the first 20 h of thermal exposure at 800 °C, due to morphological changes in the Ni thin film.

(ii) A constant degradation rate as a function of gas composition, with higher values in H_2-H_2O (0.4 % / h at $pH_2 = 3.9 \cdot 10^4$ Pa, $pH_2O = 6.2 \cdot 10^4$ Pa, T = 800 °C) than in CO-CO_2 atmosphere (0.18 % / h at pCO = $4.8 \cdot 10^4$ Pa, $pCO_2 = 5.4 \cdot 10^4$ Pa, T = 800 °C), probably due to the formation of gaseous $Ni(OH)_2$ and the diffusion of impurities to the TPB.

(iii) A slow relaxation process upon variations of pH_2 and pH_2O with settling times of 2 - 3 h for pH_2O variations and of 5 - 8 h for pH_2 variations, compared to an equilibration within few minutes in the case of CO-CO_2 atmosphere (at 800 °C).

(iv) A reproducible sequence of fast activation (few seconds) of cell performance during the recording of current-voltage curves in H_2-H_2O atmosphere, followed by a slow relaxation process with settling times of 20 h.

In order to assure reproducibility of characterization results, the following measures were taken: (i) a standardized initial heat treatment was established for all samples (20 h at T = 800 °C, $pH_2 = 8.6 \cdot 10^4$ Pa, $pH_2O = 1.5 \cdot 10^4$ Pa), (ii) a sufficient delay was applied between parameter variation and measurement (5 h), and (iii) the degradation rate was considered in data evaluation, especially at high water vapor contents. Hence, this is a significant improvement compared to previous studies on patterned Ni anodes, which did neither report degradation and relaxation processes, nor specify the entire measurement procedure.

Characterization in H_2-H_2O Atmosphere

All relevant parameter dependencies of LSR_{ct} on pH_2, pH_2O, T, and η have been determined for patterned Ni anodes by electrochemical impedance spectroscopy for parameter variation:

(i) A negative dependency of LSR_{ct} on pH_2O (b = 0.68) was found, confirming observations from literature.

(ii) The dependency on pH_2 was found to be very small.

(iii) The activation energy was found to vary with gas composition with values of E_{act} = 1.01 - 1.37 eV (T = 450 °C - 800 °C).

(iv) A charge transfer coefficient α = 0.61 - 0.69 (T = 650 - 800 °C) was determined based on impedance spectra under polarization.

Furthermore, the proportionality between LSR_{ct} and l_{TPB} has been affirmed by the characterization of samples with different design.

Additionally, the effect of a double layer capacitance at the electrode-electrolyte interface has been considered for the first time in the evaluation of the capacitive contribution of patterned Ni anodes. Herein, the separation into a Line Specific Capacitance (LSC) and an Area Specific Capacitance (ASC) was enabled by the characterization of samples with different A_{2PB}/l_{TPB} ratio. The experimentally determined values come close to the theoretical double layer capacitance calculated according to Stern theory, affirming the physical origin of the double layer.

Characterization in CO-CO_2 Atmosphere

The parameter dependencies of LSR_{ct} on pCO, pCO_2 and T have been determined for a large parameter variation under avoidance of Ni oxidation and carbon formation:

(i) A negative dependency of LSR_{ct} on pCO_2 ($d = 0.61 - 0.79$) was found.

(ii) A kink in the dependency on pCO at $pCO \approx 2{\cdot}10^4$ Pa was found: for $pCO < 2{\cdot}10^4$ Pa a negative dependency is obtained, for higher values of pCO either a weaker negative or a direct proportionality is obtained, depending on the value of pCO_2.

(iii) A change in activation energy from 0.85 – 1.42 eV is observed with increasing pCO.

Additionally, the expected proportionality between LSR_{ct} and l_{TPB} has been found, affirming that the electrochemical oxidation of CO takes place at the TPB.

Comparison to Literature Data

For the first time, a detailed comparison of the electrochemical performance of patterned Ni anodes reported in literature has been performed. This was realized by the evaluation of LSR_{ct} values and their calculation to standard conditions of a constant gas composition and different temperatures, where necessary by an extrapolation of LSR_{ct} to temperatures exceeding the experimentally performed 800 °C. The observed significant deviations between previous studies – 2 to 3 orders of magnitude for LSR_{ct} and scatter in gas composition dependencies – are attributed to (i) a disregard of the gas diffusion contribution in the analysis of impedance spectra, (ii) differences in the measurement procedure regarding relaxation and degradation processes, as well as (iii) an erroneous estimation of the active TPB length and a possible blocking of parts of the TPB with impurities.

The comparison to the electrochemical performance of pure point electrodes shows very good agreement with a deviation of less than one order of magnitude. Additionally, the high purity of the characterized patterned anodes is emphasized by the similarity in the small size of the impurity features at the TPB.

In the case of CO-CO_2 atmospere, considerably higher LSR values are reported for Ni grid anodes (deviations exceeding one order of magnitude by far). These deviations and the different parameter dependencies are assumed to be due to a considerable amount of impurities at the TPB of grid electrodes as a consequence of an annealing process at 1400 °C.

Comparison to Technical Cermet Anodes

The comparison of the electrochemical performance of patterned Ni anodes and Ni/8YSZ cermet anodes was enabled by a transmission line calculation that combines the contributions of charge transfer at the TPB and oxygen ion transport in the YSZ.

Good agreement has been obtained for the calculated ASR_{TL} and the experimentally determined ASR_{AFL} of an anode supported cell characterized by Leonide. This agreement holds especially for absolute ASR values and the activation energy in H_2-H_2O and CO-CO_2 atmosphere. For the parameter dependencies on gas composition, the transmission line calculation results in stronger dependencies than experimentally observed. This is attributed to a too high R_3/R_1 ratio in the calculation either due to (i) an erroneous estimation of the parameters $l_{TPB,v}$ and $\sigma_{ion,eff}$, (ii) an underestimation of the ionic degradation, or (iii) a slightly stronger influence of impurities at the TPB of patterned anodes than in technical cermet anodes.

Reaction Mechanisms

Similar results for (i) absolute LSR_{ct} values and (ii) parameter dependencies have been obtained for the electrochemical characterization in H_2-H_2O and CO-CO_2 atmosphere, indicating similar reaction mechanisms in both atmospheres. In H_2-H_2O atmosphere, a hydrogen spillover is considered as most probable rate determining step. Interstitial charge transfer is assumed ineffective for CO-CO_2. However, it may contribute to the current density in H_2-H_2O and be responsible for the observed slow relaxation processes.

The good agreement to the performance of cermet anodes indicates the same reaction mechanisms at the TPB of patterned anodes and cermet anodes, even though an additional limitation due to impurities at the TPB of patterned anodes cannot be excluded with respect to the uncertainties in the estimation of the transmission line parameters $l_{TPB,v}$ and $\sigma_{ion,eff}$.

Finally, the applicability of patterned anodes for the investigation of the electrochemical oxidation of the fuel gas at SOFC anodes has been confirmed under the precondition that the stability of the patterned anode as well as the relaxation and degradation processes are considered during the electrochemical characterization.

In a next step, the detailed knowledge gained by the characterization of patterned Ni anodes in H_2-H_2O and CO-CO_2 atmosphere will be extended to the analysis of the electrochemical oxidation of reformate gases at the anode of SOFCs. Herein, the comparison of the performance of patterned Ni anodes and Ni/8YSZ cermet anodes will contribute to elucidate the reaction mechanism and the reaction site of H_2 and CO oxidation respectively. Additionally, detailed modeling studies of the elementary kinetic reaction mechanism are enabled based on this work.

9 Appendix

9.1 Comparison of Sample Contacting

In this work, two different setups have been chosen for the samples with different Ni pattern design, see the setups in section 3.2, figure 3.4: (i) the samples with design #1, with two thin perpendicular stripes at the ends, are contacted by a Ni mesh on the whole surface, denoted 'Ni mesh/no flowfield' (figure 3.4a). (ii) The samples with design #2, with a 500 µm broad frame, are contacted by a special contact block, positioned on two opposite sides of the sample, which improves the gas flow to the TPB. In order to reduce the contribution of gas diffusion, the Ni sample holder was shaped with a flow-field, and an Ni mesh ensures contacting of the Ni/8YSZ cermet electrode (counter electrode), denoted 'contact block/flowfield' (figure 3.4b).

The considerable decrease in gas diffusion contribution $R_{A\alpha}$ is observed in the impedance spectrum in figure 9.1. At anodic polarization voltage, the distinction of the gas diffusion and charge transfer contribution is clearly visible. For the contacting with Ni mesh, the contribution of gas diffusion is considerable, whereas it is less pronounced for contacting with the combination of contact block and flowfield.

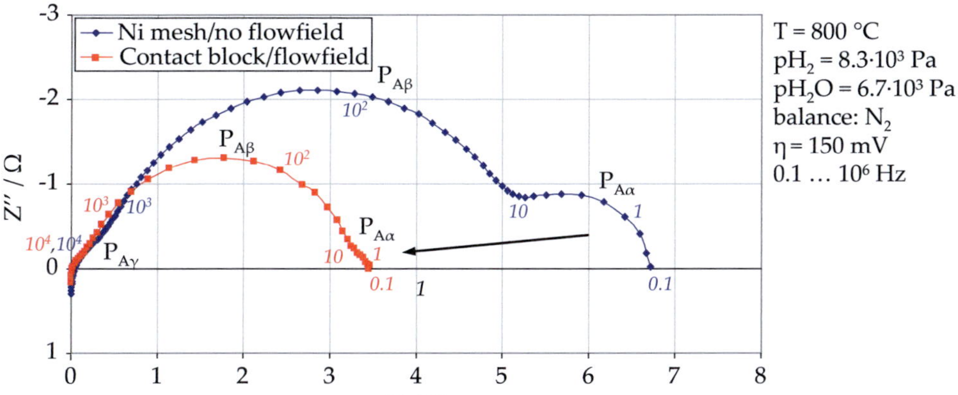

figure 9.1: Comparison of impedance spectra obtained with contacting of entire electrode surface by a Ni mesh and contacting with special contact block at same conditions with $l_{TPB,corr}$ = 6.5 m. A considerable decrease in the contribution of gas diffusion $P_{A\alpha}$ is observable. Ohmic contributions of electrolyte and measurement setup are subtracted.

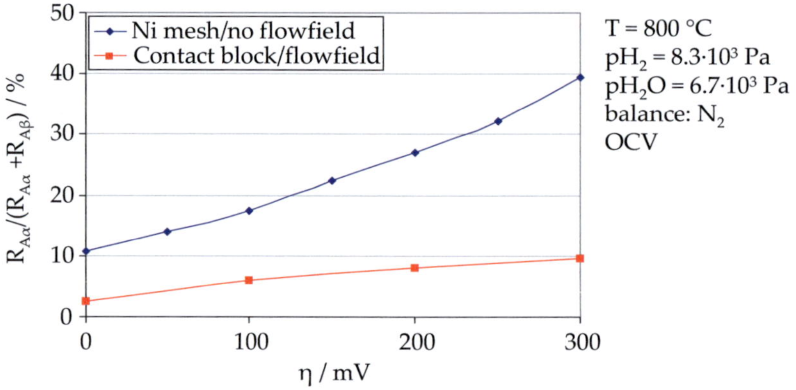

figure 9.2: Comparison of gas diffusion contribution R_α to overall contribution of the patterned Ni anode for the two different contacting methods as a function of the polarization voltage η. A strong decrease in gas diffusion contribution between contacting of the entire surface compared to the special contact block is observed.

The ratio of gas diffusion resistance to the sum of gas diffusion and charge transfer for the two different contact methods is shown in figure 9.2. The contribution of gas diffusion increases with polarization voltage since R_{ct} decreases. At 300 mV polarization, $R_{A\alpha}$ accounts for 40 % in the case of contacting with Ni mesh, whereas it accounts only for 10 % in the case of contacting with the contact block.

As expected, the contribution of charge transfer ($R_{A\beta} = R_{ct}$) is not affected by the change in contacting method, see figure 9.3: the values of $R_{A\beta}$ for the two different sample designs as a function of $l_{TPB,corr}$ are in good agreement, with respect to the strong sample to sample deviations that are observed in patterned Ni anode characterization.

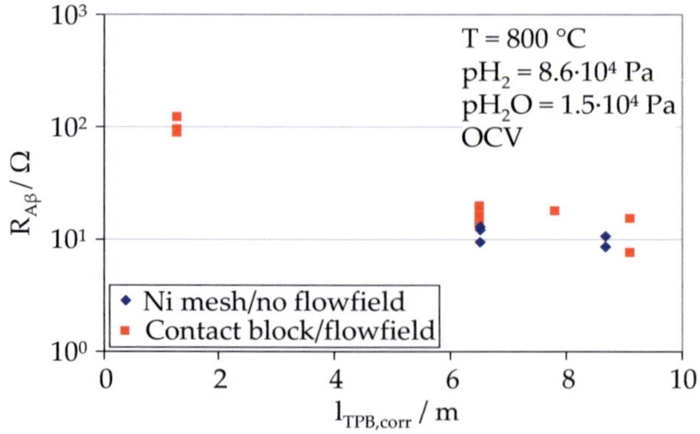

figure 9.3: Comparison of charge transfer contribution $R_{A\beta}$ for the two different contacting methods as a function of $l_{TPB,corr}$ (corrected for increase in l_{TPB} due to grain growth during initial thermal exposure). The good agreement shows that determination of charge transfer contribution is not affected by the change in contacting method.

In this work, the contributions of charge transfer ($R_{A\beta}$) and gas diffusion ($R_{A\alpha}$) can be determined separately by DRT calculations for both contacting methods. Hence, the applied contacting method does no influence characterization results. However, as the gas diffusion contribution is only slightly temperature activated, while charge transfer is strongly temperature activated, it is favorable to perform the electrochemical characterization at higher temperatures in order to allow for a precise separation of the two contributions.

9.2 Calculation of Effect of Impurity Amount

In order to get an idea of the effect of SiO_2 as a barrier for the electrochemical reaction, the possibility of the electrolyte surface being covered with a monolayer of SiO_2 is evaluated as a function of the impurity content. A similar calculation may be carried out for other impurities and the results will be similar as atom sizes are all within a factor of 2 to 3. Based on the impurity content (weight fraction), x_{imp}, of SiO_2 in the electrolyte volume (V_{YSZ}), the overall amount of SiO_2 is calculated. This amount is then related to the electrolyte surface A_{YSZ}, using the base area of the SiO_2 unit cell ($A_{SiO2,uc} = 8.8\cdot10^{-20}$ m^2). Herein, the ratio of electrolyte surface covered with SiO_2 is determined under the simplifying assumption that the whole amount of impurities segregates to the surface. It is furthermore assumed that SiO_2 forms a complete monolayer rather than to agglomerate at the TPB or grain boundaries.

The equation for the determination of the maximum area (A_{SiO2}), which a monolayer of the available SiO_2 can cover, is:

$$A_{SiO_2} = \frac{V_{YSZ} \cdot c_{YSZ} \cdot x_{imp} \cdot N_A}{M_{SiO_2}} \cdot A_{SiO_2,uc} \qquad 9.1$$

The coverage, i.e. the fraction of electrolyte surface blocked by SiO_2 is given in equation 9.2. A value higher than one indicates a coverage exceeding one monolayer of SiO_2:

$$A_{SiO_2} / A_{YSZ} = \frac{V_{YSZ}}{A_{YSZ}} \cdot \frac{c_{YSZ} \cdot x_{imp} \cdot N_A}{M_{SiO_2}} \cdot A_{SiO_2,uc} \qquad 9.2$$

Here, V_{YSZ} is the volume of the electrolyte, c_{YSZ} the density of YSZ (c_{YSZ} = 5.9 g·cm^{-3}), x_{imp} the impurity content of the electrolyte as absolute value, N_A the Avogadro number, M_{SiO2} the molar mass of SiO_2 (M_{SiO2} = 60 g/mol) and $A_{SiO2,uc}$ the ground area of the SiO_2 unit cell.

The critical impurity content $x_{imp,crit}$ for a complete coverage of the surface with one monolayer of SiO_2 is calculated by setting $A_{SiO2}/A_{YSZ} = 1$ and resolving for x_{imp}:

$$x_{imp,crit} = \frac{A_{YSZ}}{V_{YSZ}} \cdot \frac{M_{SiO_2}}{A_{SiO_2,uc} \cdot c_{YSZ} \cdot N_A} \qquad 9.3$$

From equations 9.2 and 9.3 it is evident, that the ratio of electrolyte surface to electrolyte volume is decisive for the fraction of electrolyte surface covered with SiO_2. Hence, completely different situations arise for the effect of impurities on large compact electrolytes compared to nano- or micro-scaled porous electrode-electrolyte structures: the deviation of surface to volume ratio is more than three orders of magnitude between a sphere with 1 μm diameter ($A_{YSZ}/V_{YSZ} = 6\cdot10^6$ m^{-1}) and an electrolyte with 20 x 20 x 0.5 mm^3 ($A_{YSZ}/V_{YSZ} = 4.2\cdot10^3$ m^{-1}).

The calculation of the surface to volume ratio as well as the critical impurity content for one monolayer of SiO_2 at the electrolyte surface for several electrolyte geometries used in point and patterned anode studies is given in table 9.1.

table 9.1: Geometry aspects and resulting critical impurity content for 1 monolayer of SiO_2 coverage on the electrolyte for different electrolyte designs of point and patterned electrodes.

Name	Ref.	Geometry	Width or diameter [mm]	Thickness [mm]	A_{YSZ}/V_{YSZ} [abs]	$x_{imp,crit}$ [w. %]	$x_{imp,crit}$ [ppm by mol.]	A_{SiO2}/A_{YSZ} (x_{imp}=0.0025w.%) [abs]
this work		square	20	0.2	$1.02 \cdot 10^4$	0.000195	3.9	10
this work		square	20	0.5	$4.20 \cdot 10^3$	0.000080	1.6	25
Mizusaki	[8]	round	11.3	1	$2.35 \cdot 10^3$	0.000045	0.9	44
Bieberle	[6]	round	25	0.5	$4.16 \cdot 10^3$	0.000079	1.6	25
De Boer	[7]	round	25	0.13	$1.55 \cdot 10^4$	0.000300	6.0	7
Høgh	[21]	square	10	3	$1.07 \cdot 10^3$	0.000020	0.4	97
Risø 3E	[115]	'round'	7.5 / 3	16.9 / 14.9	$9.78 \cdot 10^2$	0.000019	0.4	105

The critical impurity content that (in the very worst case that all SiO_2 ends up at the surface) theoretically may lead to a complete coverage of the electrolyte surface with SiO_2 for different geometries is shown in figure 9.4. The large compact electrolytes typically used in point and patterned anode characterization exhibit an unfavorable surface to volume ratio and imply strong requirements to impurity level. The Risø three electrode electrolyte (denoted 'Risø 3E') is already completely covered at impurity levels as low as 0.000019 w. % (0.4 ppm, refers to molar percentage). The most advantageous electrolyte design used by de Boer, requires impurity levels below 0.0003 w. % (6 ppm). This worst case estimate indicates that it is almost impossible to avoid complete coverage of the surface by using very clean electrolyte and electrode materials.

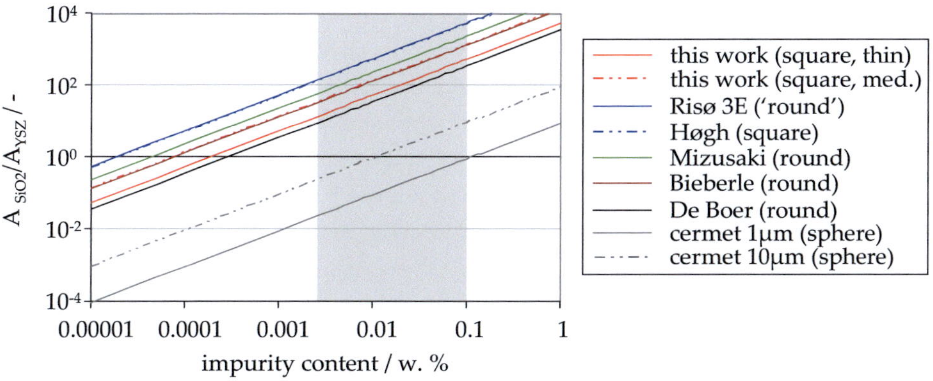

figure 9.4: Estimation of the fraction of electrolyte surface covered with one monolayer of SiO_2 as a function of the impurity content under the assumption that all impurities segregate to the surface, see equation 9.2. Different electrolyte designs for the characterization of point and patterned electrodes are given and compared to cermet structures with spheres of 1 and 10 µm diameter respectively. The grey area indicates typical impurity levels of YSZ between 0.0025 - 0.1 w. %.

With thermodynamical calculations, de Ridder et al. state that the impurity content needs to be below 10 ppm to avoid full coverage of the surface [75]. However, de Ridder did not take into account the surface to volume ratio and the absolute quantity of impurities, but seems to be only considering the low solubility of impurities in the electrolyte compared to the high free surface energy of the YSZ electrolyte.

For micro- or nano-structured electrolytes such as in the Ni/8YSZ cermet, a complete coverage of the surface would require higher impurity levels. Moreover, the maximum coverage reduces with reducing particle size at a given impurity level, see figure 9.5. For a diameter of 10 µm, one monolayer of SiO_2 is obtained at an impurity level of 0.011 w. % (220 ppm). Whereas for 1 µm particles, a value of 0.11 w. % (2200 ppm) would be required.

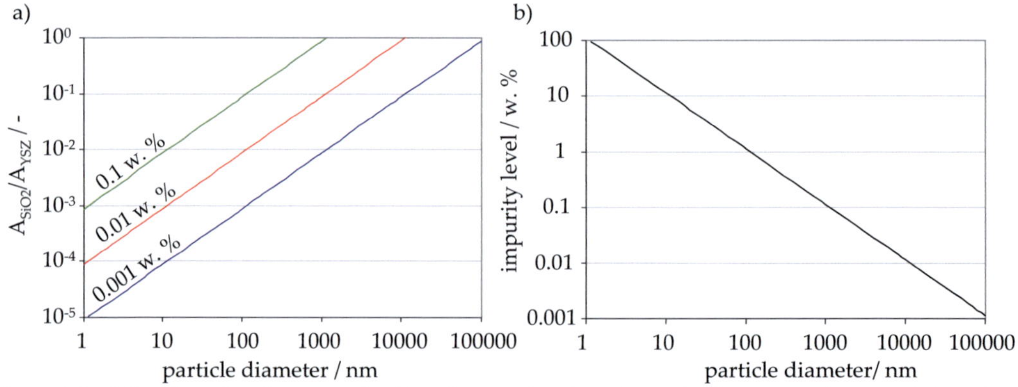

figure 9.5: a) Fraction of surface covered with SiO_2 as a function of particle diameter in the case of electrolyte spheres for three different impurity levels of 0.1, 0.01 and 0.001 w. %, and b) impurity level that a monolayer of SiO_2 constitutes of the total YSZ particle as a function of the particle diameter.

Typical impurity levels for commercial YSZ electrolytes vary strongly with values as low as 0.0025 w. % (ca 50 ppm.) [75] and up to 0.1 w. % (ca. 2000 ppm) [116]. The fraction of covered surface for the smallest impurity level of 50 ppm in the right column of table 9.1 demonstrates that it is impossible to avoid impurity coverage of the YSZ electrolyte, see also figure 9.4. Even in the case that very pure substrates are used, 7 to 105 times the electrolyte surface is covered with SiO_2, thus one monolayer is exceeded by far (always assuming that at working temperature in the range 600 – 1000 °C the majority of impurities has segregated at the interface). For cermet structures, the situation is less critical: even in the case of very impure materials with 0.1 w. %, no entire SiO_2 monolayer is forming for particles smaller than 1 µm in diameter, see figure 9.5. However, as both Ni and YSZ contain other impurities, and as the impurities tend to segregate to the TPB, there is never a guarantee of really 'clean' TPBs in any electrode, if 'clean' is taken as the surfaces at the TPB that have the same composition as the bulk.

9.3 Estimation of TPB Length

In table 9.2, the values or TPB length reported in literature and determined following two different methods, namely 3D reconstruction and microstructure modeling, are summarized.

*table 9.2: Comparison of values of l_{TPB} reported in literature together with the percolation percentage and the penetration depth. In the second column from the right, the resulting l_{TPB} value, including percolation and penetration depth is given in m·cm^{-2}. The subscript v refers to volumetric $l_{TPB,v}$ values in m·m^{-3}. * The penetration depth is adopted from a study by Brown et al. [32].*

Name	Ref.	YSZ-Ni-pore [vol %]	$l_{TPB,v}$ [m^{-2}]	Perc. [%]	$l_{TPB,per,v}$ [m^{-2}]	λ [μm]	l_{TPB} [m·cm^{-2}]	Comment
Wilson	[104]	55-26-19	$4.3 \cdot 10^{12}$	63	$2.7 \cdot 10^{12}$	Ca. 10 *	2700	3D reconstruction
Shearing	[103]	20-72-8	$9.4 \cdot 10^{12}$	45	$4.2 \cdot 10^{12}$	n.a.	-	3D reconstruction
		27-63-9	$1.1 \cdot 10^{13}$	51	$5.4 \cdot 10^{12}$	n.a.	-	
Thyden	[117]	55-25-20	n.a.	60	n.a.	2D plane	53	3D reconstruction, value for electrode-electrolyte interface
Gewies	[39]	34.5-30.5-35			$4.0 \cdot 10^{12}$			Microstructure model, percolation probabilities
This work		35-30-35	$1.1 \cdot 10^{12}$ - $4.4 \cdot 10^{12}$	98	$1.0 \cdot 10^{12}$ - $4.3 \cdot 10^{12}$	3 to 6	300 - 2600	Microstructure model and simulation of performance
		48-30-22	$9.5 \cdot 10^{11}$ - $3.8 \cdot 10^{12}$	96	$9.0 \cdot 10^{11}$ - $3.6 \cdot 10^{12}$	4 to 6		

9.4 List of Samples

Cell #	A [cm²]	Ni stripe [μm]	Spacing [μm]	$l_{TPB,th}$ [m·cm^{-2}]	$l_{TPB,corr}$ [m·cm^{-2}]	Frame [μm]	Electrolyte	Comment
NiYSZ_139	1^2	15	15	6.7	8.7	15	8.5 YSZ, PC	H_2-H_2O, Al_2O_3 holder
NiYSZ_151	1^2	15	15	6.7	8.7	15	8.5 YSZ, PC	H_2-H_2O, Al_2O_3 holder
NiYSZ_143	1^2	15	15	6.7	8.7	15	8.5 YSZ, PC	H_2-H_2O, Al_2O_3 holder
NiYSZ_146	1^2	15	15	6.7	8.7	15	8.5 YSZ, PC	H_2-H_2O, Al_2O_3 holder
NiYSZ_140	1^2	20	20	5.0	6.5	20	8.5 YSZ, PC	H_2-H_2O, Al_2O_3 holder
NiYSZ_148	1^2	20	20	5.0	6.5	20	8.5 YSZ, PC	H_2-H_2O, Al_2O_3 holder
NiYSZ_150	1^2	20	20	5.0	6.5	20	8.5 YSZ, PC	H_2-H_2O, Al_2O_3 holder
NiYSZ_141	1^2	20	20	5.0	6.5	20	8.5 YSZ, PC	H_2-H_2O, Al_2O_3 holder
NiYSZ_211	1.025^2	20	15	5.0	6.5	500	8.5 YSZ, PC	H_2-H_2O, Ni holder
NiYSZ_217	1.025^2	20	15	5.0	6.5	500	8.5 YSZ, PC	H_2-H_2O, Ni holder
NiYSZ_256	1.025^2	20	15	5.0	6.5	500	8.5 YSZ, PC	H_2-H_2O, Ni holder
NiYSZ_220	1.025^2	12	10	8.0	10.4	500	8.5 YSZ, PC	H_2-H_2O, Ni holder
NiYSZ_223	1.025^2	12	13	7.0	9.1	500	8.5 YSZ, PC	H_2-H_2O, CO-CO_2, Ni holder
NiYSZ_225	1.025^2	12	13	7.0	9.1	500	8.5 YSZ, PC	H_2-H_2O, CO-CO_2, Ni holder
NiYSZ_246	1.025^2	20	9	6.0	7.8	500	8.5 YSZ, PC	H_2-H_2O, CO-CO_2, Ni holder
NiYSZ_280	1.025^2	25	160	1.0	1.3	500	8.5 YSZ, PC	H_2-H_2O, Ni holder
NiYSZ_259	1.025^2	25	160	1.0	1.3	500	8.5 YSZ, PC	CO-CO_2, Ni holder
NiYSZ_291	1.025^2	25	160	1.0	1.3	500	8.5 YSZ, PC	CO-CO_2, Ni holder
NiYSZ_276	1.025^2	20	15	5.0	6.5	500	8.5 YSZ, PC	CO-CO_2, Ni holder
DLC_1	1^2	-	-	0.04	0.4	-	8.5 YSZ, PC	H_2-H_2O, Al_2O_3 holder

9.5 Symbols

α	charge transfer coefficient, for $n_e = 2$ [-]
α'	charge transfer coefficient, for $n_e = 1$ [-]
β_{rel}	relative water vapor content [-], [%]
$\beta_{rel,CO/CO2}$	pCO_2/pCO ratio according to relative water vapor content [-], [%]
ε	porosity [-]
ε_0	vacuum permittivity [$\varepsilon_0 = 8.854 \cdot 10^{-12}$ F·m^{-1}]
ε_r	dielectric constant / relative permittivity [-]
ζ	interface impedance at TPB [Ω·m]
η	polarization voltage [V]
η_{ct}	polarization voltage attributed to charge transfer [V]
Θ_H	surface coverage with hydrogen [-]
λ	penetration depth for transmission line model [m]
σ_{el}	electronic conductivity [S·m^{-1}]
σ_{ion}	ionic conductivity [S·m^{-1}]
$\sigma_{ion,bulk}$	ionic conductivity of bulk material [S·m^{-1}]
$\sigma_{ion,bulk,d}$	degraded ionic conductivity of bulk material [S·m^{-1}]
$\sigma_{ion,eff}$	effective ionic conductivity [S·m^{-1}]
$\sigma_{ion,eff,d}$	degraded effective ionic conductivity [S·m^{-1}]
τ	tortuosity [-]
τ_{corr}	correction factor [-]
χ_1	resistance load attributed to ionic transport in YSZ [Ω·m^{-1}]
χ_2	resistance load attributed to electronic transport in Nickel [Ω·m^{-1}]
A	area [m^2]
A_{2PB}	area of 2PB (electrode-electrolyte interface) [m^2]
$A_{SiO2,uc}$	base area of SiO_2 unit cell [$A_{SiO2,uc} = 8.8 \cdot 10^{-20}$ m^2]
a	exponent for pH_2 dependency of j_0 and LSR [-]
ASR	Area Specific Resistance [Ω·cm^2]
ASR_{AFL}	Area Specific Resistance of anode functional layer [Ω·cm^2]
ASR_{ESC}	Area Specific Resistance of electrolyte supported cell [Ω·cm^2]
ASR_{ct}	ASR of charge transfer [Ω·cm^2]
ASR_{TL}	ASR of charge transfer of transmission line model [Ω·cm^2]
$ASR_{TL,d}$	ASR of charge transfer of transmission line with degrad. ionic conductivity [Ω·cm^2]
b	exponent for pH_2O dependency of j_0 and LSR [-]
c	exponent for pCO dependency of j_0 and LSR [-]
	integration constant [-]
c_i	proportionality factor, with subscript i = 1, 2, 3, 4, 5, 6, 7, 8
c_{YSZ}	density of YSZ unit cell [$c_{YSZ} = 2.98 \cdot 10^{28}$ m^{-3}]
C_{ct}	capacitance attributed to charge transfer [F]

C_{diff}	diffuse capacitance [$F \cdot m^{-2}$]
C_{dl}	capacitance attributed to double layer [F]
C_{Hh}	internal Helmholtz double layer [F]
d	exponent for pCO_2 dependency of j_0 and LSR [-]
d_{O2-}	diameter of O^{2-} [d_{O2-} = 140 pm]
e	elementary charge [$e = 1.6022 \cdot 10^{-19}$ C]
E_{act}	activation energy [eV]
$E_{act,app}$	apparent activation energy (including charge transfer and ionic transport) [eV]
$E_{act,R1}$	activation energy attributed to R_1 [eV]
$E_{act,R3}$	activation energy attributed to R_3 [eV]
$E_{act,RTL}$	activation energy attributed to R_{TL} [eV]
f	structural parameter [-]
f_{rel}	relaxation frequency [Hz]
F	Faraday constant [$F = 96485.3\ C \cdot mol^{-1}$]
$G_{0,H2-H2O}$	Gibbs free energy of hydrogen oxidation reaction [$kJ \cdot mol^{-1}$]
j	current density [$A \cdot cm^{-2}$]
j_0	exchange current density [$A \cdot cm^{-2}$]
k_B	Boltzmann constant [$k = 8.617 \cdot 10^{-5}\ eV \cdot K^{-1}$]
l_{TPB}	TPB length [$m \cdot cm^{-2}$]
$l_{TPB,corr}$	corrected TPB length, considering morphological changes [$m \cdot cm^{-2}$]
$l_{TPB,th}$	theoretical TPB length, based on pattern design [$m \cdot cm^{-2}$]
$l_{TPB,v}$	volumetic TPB length [$m \cdot m^{-3}$]
L	length of transmission line [m]
	inductance of measurement setup [H]
LSR	line specific resistance [$\Omega \cdot m$]
LSR_{ct}	line specific resistance attributed to charge transfer [$\Omega \cdot m$]
M_{SiO2}	molar mass of SiO_2 [$M_{SiO2} = 60\ g \cdot mol^{-1}$]
n	exponent of RQ element, in the case of RC element n = 1 [-]
N_A	Avogadro number [$N_A = 6.0221 \cdot 10^{23}\ mol^{-1}$]
n_e	number of transferred electrons (for SOFC $n_e = 2$)
$P_{A\alpha}$	process of impedance spectrum: gas diffusion
$P_{A\beta}$	process of impedance spectrum: charge transfer at patterned Ni anode
$P_{A\gamma}$	process of impedance spectrum: charge transfer at counter anode
p_i	partial pressure [Pa]
pCO	partial pressure of carbon monoxide [Pa]
pCO_2	partial pressure of carbon dioxide [Pa]
pH_2	partial pressure of hydrogen [Pa]
$pH_{2,m}$	measured partial pressure of hydrogen [Pa]
$pH_{2,s}$	standard partial pressure of hydrogen [here: $pH_{2,s} = 10^4$ Pa]
pH_2O	partial pressure of water vapor [Pa]

pN_2	partial pressure of nitrogen [Pa]
pO_2	partial pressure of oxygen [Pa]
R	universal gas constant [$R = 8.3145\ J{\cdot}mol^{-1}{\cdot}K^{-1}$]
	resistance
r_1	resistance load attributed to ionic transport in YSZ [$\Omega{\cdot}m^{-1}$]
r_2	resistance load attributed to electronic transport in Nickel [$\Omega{\cdot}m^{-1}$]
r_3	distributed DC interface resistance at TPB [$\Omega{\cdot}m$]
R_1	total DC resistance of ionic transport of transmission line [Ω]
$R_{1,CA}$	total DC resistance of ionic transport of TL for cermet anode [Ω]
$R_{1,PA}$	total DC resistance of ionic transport of TL for patterned anode [Ω]
$R_{1,d,PA}$	total DC resistance of degrad. ionic transport of TL for patterned anode [Ω]
R_3	total DC resistance of charge transfer at TPB [Ω]
$R_{3,ESC}$	total DC resistance of charge transfer at TPB of electrolyte supported cell [Ω]
$R_{3,PA}$	total DC resistance of charge transfer at TPB of patterned anode [Ω]
$R_{3,l,PA}$	total DC resistance of CT at TPB with adapted l_{TPB} of patterned anode [Ω]
R_{ct}	polarization resistance attributed to charge transfer [Ω]
R_{ohm}	ohmic resistance of the electrolyte [Ω]
R_{pol}	polarization resistance [Ω]
R_{setup}	ohmic resistance of the measurement setup [Ω]
R_{TL}	polarization resistance of transmission line
T	temperature [K], [°C]
U_{Nernst}	Nernst voltage [V]
$[V_O^{\bullet\bullet}]_{bulk}$	molar fraction of oxygen vacancies in the YSZ bulk [-]
x_{H2O}	molar fraction of H_2O [-]
x_{imp}	impurity content of SiO_2 in weight fraction [-]
z_1	complexe resistance load attributed to ionic transport in YSZ [$\Omega{\cdot}m^{-1}$]
z_2	complexe resistance load attributed to electronic transport in Nickel [$\Omega{\cdot}m^{-1}$]
z_3	complexe interface impedance at TPB [$\Omega{\cdot}m$]
Z	complexe impedance [Ω]
Z'	real part of the complexe impedance [Ω]
Z''	imaginary part of the complexe impedance [Ω]
Z_{TL}	complexe impedance [Ω]

9.6 Abbreviations

2PB	Two-Phase-Boundary (electrode-electrolyte interface)
8YSZ	8.5 mol% Y_2O_3 stabilized ZrO_2 → see YSZ
at. %	atomic percent
AFL	Anode Functional Layer
AFM	Atomic Force Microscopy
an	anode, anodic
ASC	Anode Supported Cell
	Area Specific Capacitance
ASR	Area Specific Resistance
cat	cathode, cathodic
ct	charge transfer
C	Capacitance
CE	Cermet Electrode / Counter Electrode
CNLS	Complex Nonlinear Least Squares
CPE	Constant Phase Element
CT	Charge Transfer
C/V	Current-Voltage curve
dl	double layer
DRT	Distribution of Relaxation Times
EDX	Energy Dispersive X-ray Diffraction
EEI	Electrode-Electrolyte Interface
EELS	Electron Energy Loss Spectroscopy
EIS	Electrochemical Impedance Spectrum
ESC	Electrolyte Supported Cell
FRA	Frequency Response Analyser
IWE	Institute of Materials for Electrical and Electronic Engineering, at Karlsruhe Institute of Technology (KIT)
LSC	Line Specific Capacitance
LSR	Line Specific Resistance
MALT	Materials Orientated Little Thermodynamic Database
MFC	Mass Flow Controller
Ni/8YSZ	Ni-8YSZ cermet structure
OCV	Open Circuit Voltage
ppm	parts per million
rpm	rotations per minute
R	Resistor
RC	Resistor R in parallel with a capacitance C

RQ	Resistor R in parallel with a constant phase element Q
ScYSZ	Scandia and Yttria stabilized Zr_2O_3
SEM	Scanning Electron Microscopy
SOFC	Solid Oxide Fuel Cell
SPIP	Scanning Probe Image Processor
TPB	Triple-Phase-Boundary
TPD	Temperature Programmed Desorption
TL	Transmission Line
TOF-SIMS	Time-of-Flight Secondary Ion Mass Spectroscopy
w. %	weight percent
YSZ	Y_2O_3 stabilized ZrO_2
XPS	X-ray Photoelectron Spectrometry

9.7 List of Figures

9.8 List of Tables

9.9 Supervised Diploma Thesis and Study Projects

- Jean-Claude Njodzefon, Stabilitätsuntersuchungen an Nickelstrukturen, Study Project, IWE, Universität Karlsruhe (TH), 2008.
- Thassilo Knapp, Herstellung von Modellanoden durch physikalisches Ätzen, Diploma Thesis, IWE, Universität Karlsruhe (TH), 2009.
- Lisa Braun, Bestimmung der Doppelschichtkapazität für Modellanoden, Bachelor Thesis, IWE, Karlsruher Institut für Technologie (KIT), 2010

9.10 Own Publications

9.10.1 Publications

(1) A. Utz, J. Joos, A. Weber and E. Ivers-Tiffée, 'Electrochemical Oxidation at SOFC Anodes: Comparison of Patterned Ni Anodes and Technical Ni/8YSZ Cermet Anodes', in *ECS Transactions*, submitted 12/2010.

(2) A. Utz, K.V. Hansen, K. Norrman, E. Ivers-Tiffée and M. Mogensen, 'Impurity Features in Ni-YSZ-H2-H2O Electrodes', in *Solid State Ionics* 10.1016/j.ssi.2010.12.014.

(3) A. Utz, A. Leonide, A. Weber and E. Ivers-Tiffée, 'Studying the CO/CO2 Characteristics of SOFC Anodes by Means of Patterned Ni Anodes' in *Journal of Power Sources* 10.1016/j.jpowsour.2010.10.056.

(4) A. Utz, H. Störmer, D. Gerthsen, A. Weber and E. Ivers-Tiffée, 'Microstructure Stability Studies of Ni Patterned Anodes for SOFC', in *Solid State Ionics* 10.1016/j.ssi.2010.05.004.

(5) W. Bessler, M. Vogler, H. Störmer, D. Gerthsen, A. Utz, A. Weber, E. Ivers-Tiffée, 'Model Anodes and Anode Models for Understanding the Mechanism of Hydrogen Oxidation in Solid Oxide Fuel Cells', in *Physical Chemistry Chemical Physics* 12 (2010) p. 13888-13903.

(6) A. Utz, A. Leonide, A. Weber and E. Ivers-Tiffée, 'Studying the CO/CO_2 Characteristics of SOFC Anodes by Means of Ni Patterned Anodes' in *Proceedings of the 9th European Solide Oxide Fuel Cell Forum* (2010) p. 9.55-9.70.

(7) A. Utz, T. Knapp, H. Störmer, T. Knapp, K. Ilin, P.-J. Jakobs, M. Siegel and E. Ivers-Tiffée, 'A Comparative Study of Fabrication Procedures for Ni Patterned Anodes' in *Proceedings of the 9th European Solide Oxide Fuel Cell Forum* (2010) p. 5.91-5.103.

(8) M. Vogler, A. Utz, H. Störmer, A. Weber, D. Gerthsen, E. Ivers-Tiffée and W.G. Bessler, 'Electrochemistry and Mechanism of Hydrogen Oxidation at Ni/YSZ Patterned Anodes' in *Proceedings of the 9th European Solide Oxide Fuel Cell Forum* (2010) p. 9.94-9.103.

(9) H. Störmer, A. Utz, E. Ivers-Tiffée and D. Gerthsen, 'Patterned Nickel Anodes for SOFC: Microstructure and Impurity Analysis' in *Proceedings of the 9th European Solide Oxide Fuel Cell Forum* (2010) p. 5.122-5.127.

(10) A. Utz, H. Störmer, A. Leonide, A. Weber and E. Ivers-Tiffée, 'Degradation and Relaxation Effects of Ni Patterned Anodes in H2/H2O Atmosphere' in *Journal of the Electrochemical Society*, 157 (2010) B920-B930.

(11) A. Utz, H. Störmer, A. Leonide, A. Weber and E. Ivers-Tiffée, 'Degradation Effects of Ni Patterned Anodes in H2/H2O Atmosphere' in *ECS Transactions*, 25 (2009) 2013-2021.

(12) A. Utz, A. Weber, E. Ivers-Tiffée and H. Störmer, 'Studies on Pattern Anode Stability for SOFC' in *Proceedings of the 8th European Solide Oxide Fuel Cell Forum* (2008) A0509-519.

9.10.2 Conference Contributions

(13) A. Utz, H. Störmer, A. Weber, E. Ivers-Tiffée, 'Electrochemical Characterization of Ni Patterned Anodes in H2/H2O Atmosphere', *1st International Conference on Materials for Energy* (Karlsruhe, Germany), 04.07. - 08.07.2010

(14) A. Utz, A. Leonide, A. Weber, E. Ivers-Tiffée, 'Studying the CO/CO2 Characteristics of SOFC Anodes by Means of Ni Patterned Anodes', *9th EUROPEAN SOFC FORUM* (Lucerne, Switzerland), 29.06. - 02.07.2010

(15) A. Utz, H. Störmer, T. Knapp, K. Ilin, P.-J. Jakobs, M. Siegel, E. Ivers-Tiffée, 'A Comparative Study of Fabrication Procedures for Patterned Anodes', *9th EUROPEAN SOFC FORUM* (Lucerne, Switzerland), 29.06. - 02.07.2010.

(16) H. Störmer, A. Utz, E. Ivers-Tiffée, D. Gerthsen, 'Patterned Nickel Anodes for SOFC: Microstructure and Impurity Analysis', *9th EUROPEAN SOFC FORUM* (Lucerne, Switzerland), 29.06. - 02.07.2010.

(17) M. Vogler, A. Utz, H. Störmer, A. Weber, D. Gerthsen, E. Ivers-Tiffée, W. Bessler, 'Electrochemistry and Mechanism of Hydrogen Oxidation at Ni/YSZ Patterned Anodes', *9th EUROPEAN SOFC FORUM* (Lucerne, Switzerland), 29.06.-02.07.2010.

(18) A. Utz, H. Störmer, A. Weber, E. Ivers-Tiffée, 'Study on Ni Patterned Anode Reaction Kinetics by Detailed Analysis of the Activation Energy', *Electroceramics XII* (Trondheim, Norway), 13.06. - 16.06.2010.

(19) H. Störmer, A. Utz, E. Ivers-Tiffée, D. Gerthsen, 'Microstructure and Impurity Analysis of Patterned Nickel Anodes for SOFC', *Electroceramics XII* (Trondheim, Norway), 13.06. - 16.06.2010.

(20) A. Utz, H. Störmer, A. Weber, E. Ivers-Tiffée, 'Patterned Ni Anodes on YSZ: a Promising Method to Determine Reaction Kinetics at the TPB of Electrolyte-Anode-Gas', *Bunsentagung* (Bielefeld, Germany), 13.05.-15.05.2010.

(21) M. Vogler, A. Utz, H. Störmer, A. Weber, D. Gerthsen, E. Ivers-Tiffée, W.G. Bessler, 'Combined Modeling and Experimental Study of Hydrogen Oxidation at Ni/YSZ Patterned Anodes Based on Elementary Kinetics and Transport', *7th Symposium on Fuel Cell Modelling and Experimental Validation* (Lausanne, Switzerland), 23.03. - 24.03.2010.

(22) A. Utz, H. Störmer, A. Leonide, A. Weber, E. Ivers-Tiffée, 'Degradation Effects of Ni Patterned Anodes in H2/H2O Atmosphere', *SOFC XI* (Vienna, Austria), 04.10. - 09.10.2009

(23) A. Utz, V. Sonn, H. Störmer, D. Gerthsen, A. Weber, E. Ivers-Tiffée, 'A Detailed Study of Hydrogen Oxidation Reaction based on Patterned Ni Anodes for SOFC', *17th International Conference on Solid State Ionics (SSI-17)* (Toronto, Canada), 28.06. - 03.07.2009

(24) A. Utz, H. Störmer, D. Gerthsen, A. Weber, E. Ivers-Tiffée, 'Microstructure Stability Studies of Ni Patterned Anodes for SOFC', *17th International Conference on Solid State Ionics (SSI-17)* (Toronto, Canada), 28.06. - 03.07.2009.

(25) A. Utz, V. Sonn, A. Leonide, H. Störmer, D. Gerthsen, A. Weber, E. Ivers-Tiffée, 'Ni/YSZ Patterned Anodes: Microstructure and Electrochemical Properties', *Bunsentagung* (Cologe, Germany), 21.05. - 23.05.2009.

(26) A. Utz, A. Weber, E. Ivers-Tiffée, H. Störmer, D. Gerthsen, D. Starukhin, V. Yurkiv, H.-R. Volpp, M. Vogler, W.G. Bessler, 'Mechanisms of the Electrooxidation of Hydrogen on Ni/YSZ Patterned Anodes', *Bunsentagung* (Cologne, Germany), 21.05. - 23.05.2009

(27) A. Utz, M. Vogler, W.G. Bessler, A. Weber, E. Ivers-Tiffée, 'Electrochemical Characterization and Modelling of Ni/YSZ Patterned Anodes', *6th Symposium on Fuel Cell Modelling and Experimental Validation* (Bad Herrenalb, Germany), 25.03. - 26.03.2009

(28) A. Utz, A. Weber, E. Ivers-Tiffée, H. Störmer, 'Studies on Pattern Anode Stability for SOFC', *8th EUROPEAN SOFC FORUM* (Lucerne, Schweiz), 30.06. - 04.07.2008.

10 References

[1] N. Q. Minh, 'Ceramic Fuel-Cells', in *Journal of the American Ceramic Society* 76-3, p. 563-588, 1993.

[2] C. Endler, A. Leonide, A. Weber, F. Tietz and E. Ivers-Tiffee, 'Time-Dependent Electrode Performance Changes in Intermediate Temperature Solid Oxide Fuel Cells', in *Journal of the Electrochemical Society* 157-2, p. B292-B298, 2010.

[3] A. Atkinson, S. A. Barnett, R. J. Gorte, J. T. S. Irvine, A. J. McEvoy, M. Mogensen and S. C. Singhal, 'Advanced Anodes for High-Temperature Fuel Cells', in *nature materials* 3, p. 17-27, 2004.

[4] S. B. Adler und W. G. Bessler, 'Elementary Kinetic Modeling of Solid Oxide Fuel Cell Electrode Reactions', in *Handbook of Fuel Cells - Fundamentals, Technology and Applications,* John Wiley & Sons Ltd, Chichester, p. 441-462, 2009.

[5] A. Bieberle, 'The Electrochemistry of Solid Oxide Fuel Cell Anodes: Experiments, Modeling, and Simulations', Dissertation, Eidgenössische Technische Hochschule Zürich, 2000.

[6] A. Bieberle, L. P. Meier and L. J. Gauckler, 'The Electrochemistry of Ni Pattern Anodes Used As Solid Oxide Fuel Cell Model Electrodes', in *Journal of the Electrochemical Society* 148-6, p. A646-A656, 2001.

[7] B. de Boer, 'SOFC Anodes: Hydrogen Oxidation at Porous Nickel and Nickel/Yttriastabilised Zirconia Cermet Electrodes', PhD Thesis, University of Twente, The Netherlands, 1998.

[8] J. Mizusaki, H. Tagawa, T. Saito, T. Yamamura, K. Kamitani, K. Hirano, S. Ehara, T. Takagi, T. Hikita, M. Ippommatsu, S. Nakagawa and K. Hashimoto, 'Kinetic-Studies of the Reaction at the Nickel Pattern Electrode on YSZ in H_2-H_2O Atmospheres', in *Solid State Ionics* 70, p. 52-58, 1994.

[9] J. Mizusaki, H. Tagawa, T. Saito, K. Kamitani, T. Yamamura, K. Hirano, S. Ehara, T. Takagi, T. Hikita, M. Ippommatsu, S. Nakagawa and K. Hashimoto, 'Preparation of Nickel Pattern Electrodes on YSZ and Their Electrochemical Properties in H2-H2O Atmospheres', in *Journal of the Electrochemical Society* 141-8, p. 2129-2134, 1994.

[10] S. C. Singhal and K. Kendall, *High Temperature Solid Oxide Fuel Cells*, Elsevier Ltd., New York, 2003.

[11] A. C. Müller, 'Mehrschicht-Anode für die Hochtemperatur-Brennstoffzelle (SOFC)', Dissertation, Universität Karlsruhe (TH), 2004.

[12] D. Stöver, H. P. Buchkremer und J. P. P. Huijsmans, 'MEA/Cell Preparation Methods: Europe/USA', in *Handbook of Fuel Cells*, John Wiley & Sons Ltd., Chichester, UK, p. 1015, 2003.

[13] W. G. Bessler, M. Vogler, H. Störmer, D. Gerthsen, A. Utz, A. Weber and E. Ivers-Tiffée, 'Model Anodes and Anode Models for Understanding the Mechanism of Hydrogen Oxidation in Solid Oxide Fuel Cells', in *Physical Chemistry Chemical Physics* 10.1039/C0CP00541J-12, p. 13888-13903, 2010.

[14] G. O. Lauvstad, R. Tunold and S. Sunde, 'Electrochemical Oxidation of CO on Pt and Ni Point Electrodes in Contact With an Yttria-Stabilized Zirconia Electrolyte - I. Modeling of Steady-State and Impedance Behavior', in *Journal of the Electrochemical Society* 149-12, p. E497-E505, 2002.

[15] W. G. Bessler, 'Electrochemistry and Transport in Solid Oxide Fuel Cells', Habilitiationsschrift, Ruprecht-Karls-Universität Heidelberg, Germany, 2007.

[16] M. Vogler, A. Bieberle-Hutter, L. Gauckler, J. Warnatz and W. G. Bessler, 'Modelling Study of Surface Reactions, Diffusion, and Spillover at a Ni/YSZ Patterned Anode', in *Journal of the Electrochemical Society* 156-5, p. B663-B672, 2009.

[17] W. C. Conner and J. L. Falconer, 'Spillover in Heterogeneous Catalysis', in *Chemical Reviews* 95-3, p. 759-788, 1995.

[18] M. Mogensen, J. Hogh, K. V. Hansen and T. Jacobsen, 'A Critical Review of Models of the H-2/H2O/Ni/SZ Electrode Kinetics', in *ECS Transactions* 7-1, p. 1329-1338, 2007.

[19] R. J. Aaberg, R. Tunold, M. Mogensen, R. W. Berg and R. Odegard, 'Morphological Changes at the Interface of the Nickel-Yttria Stabilized Zirconia Point Electrode', in *Journal of the Electrochemical Society* 145-7, p. 2244-2252, 1998.

[20] J. Guindet, C. Roux and A. Hammou, 'Hydrogen Oxidation at the Ni/Zirconia Electrode', in *Proceedings of the Second International Symposium on Solid Oxide Fuel Cells (SOFC-II)*, p. 553-559, 1991.

[21] J. Hogh, K. V. Hansen and M. Mogensen, 'Atmosphere Variations at the H2/H2O/Ni/YSZ Point Electrode at 700 °C', in *Solid State Electrochemistry*, p. 235-247, 2005.

[22] K. V. Jensen, S. Primdahl, I. Chorkendorff and M. Mogensen, 'Microstructural and Chemical Changes at the Ni/YSZ Interface', in *Solid State Ionics* 144-3-4, p. 197-209, 2001.

[23] K. V. Jensen, R. Wallenberg, I. Chorkendorff and M. Mogensen, 'Effect of Impurities on Structural and Electrochemical Properties of the Ni-YSZ Interface', in *Solid State Ionics* 160-1-2, p. 27-37, 2003.

[24] D. Kek, M. Mogensen and S. Pejovnik, 'A Study of Metal (Ni, Pt, Au)/Yttria-Stabilized Zirconia Interface in Hydrogen Atmosphere at Elevated Temperature', in *Journal of the Electrochemical Society* 148-8, p. A878-A886, 2001.

[25] T. Norby, O. J. Velle, H. Leth-Olsen and R. Tunold, 'Reaction Resistance in Relation to Three Phase Boundary Length of Ni/YSZ Electrodes', in *Proceedings of the Third International Symposium on Solid Oxide Fuel Cells (SOFC-III)*, p. 473-477, 1993.

[26] M. S. Schmidt, K. V. Hansen, K. Norrman and M. Mogensen, 'Characterisation of the Ni/ScYSZ Interface in a Model Solid Oxide Fuel Cell Anode', in *Solid State Ionics* 179-39, p. 2290-2298, 2008.

[27] A. Ehn, J. Hogh, M. Graczyk, K. Norrman, L. Montelius, M. Linne and M. Mogensen, 'Electrochemical Measurements Made in Dry CO/CO2 and H2/H2O Atmospheres on Single Crystal YSZ Using Pattern Electrodes', in *Solid State Electrochemistry*, p. 177-184, 2005.

[28] J. Mizusaki, T. Yamamura, N. Mori, H. Tagawa, K. Hirano, S. Ehara, T. Takagi, M. Hishinuma, H. Sasaki, T. Sogi, Y. Nakamura and K. Hashimoto, 'Kinetic Studies on Ni/YSZ Anode Reaction of SOFC in H2-H2O Atmospheres by the Use of Nickel Pattern Electrodes', in *High Temperature Electrochemistry: Ceramics and Metals*, p. 363-368, 1996.

[29] A. M. Sukeshini, B. Habibzadeh, B. P. Becker, C. A. Stoltz, B. W. Eichhorn and G. S. Jackson, 'Electrochemical Oxidation of H-2, CO, and CO/H-2 Mixtures on Patterned Ni Anodes on YSZ Electrolytes', in *Journal of the Electrochemical Society* 153-4, p. A705-A715, 2006.

[30] T. Yamamura, H. Tagawa, T. Saito, J. Mizusaki, K. Kamitani, K. Hirano, S. Ehara, T. Takagi, Y. Hishinuma, H. Sasaki, T. Sogi, Y. Nakamura and K. Hashimoto, 'Reaction Kinetics at the Nickel Pattern Electrode on YSZ and Its Dependence on Temperature', in *Proceedings of the Fourth International Symposium on Solid Oxide Fuel Cells (SOFC-IV)*, p. 741-749, 1995.

[31] R. Barfod, A. Hagen, S. Ramousse, P. V. Hendriksen and M. Mogensen, 'Break Down of Losses in Thin Electrolyte SOFCs', in *Fuel Cells* 6-2, p. 141-145, 2006.

[32] M. Brown, S. Primdahl and M. Mogensen, 'Structure/Performance Relations for Ni/Yttria-Stabilized Zirconia Anodes for Solid Oxide Fuel Cells', in *Journal of the Electrochemical Society* 147-2, p. 475-485, 2000.

[33] A. Leonide, V. Sonn, A. Weber and E. Ivers-Tiffée, 'Evaluation and Modeling of the Cell Resistance in Anode-Supported Solid Oxide Fuel Cells', in *Journal of the Electrochemical Society* 155-1, p. B36-B41, 2008.

[34] M. Mogensen and S. Skaarup, 'Kinetic and Geometric Aspects of Solid Oxide Fuel Cell Electrodes', in *Solid State Ionics* 86-8, p. 1151-1160, 1996.

[35] V. Sonn, A. Leonide and E. Ivers-Tiffée, 'Combined Deconvolution and CNLS Fitting Approach Applied on the Impedance Response of Technical Ni/8YSZ Cermet Electrodes', in *Journal of the Electrochemical Society* 155-7, p. B675-B679, 2008.

[36] A. Ehn, J. Hogh, M. Graczyk, K. Norrman, L. Montelius, M. Linne and M. Mogensen, 'Electrochemical Investigation of Nickel Pattern Electrodes in H2/H2O and CO/CO2 Atmospheres', in *Journal of the Electrochemical Society* 157-11, p. B1588-B1596, 2010.

[37] M. V. Rao, J. Fleig, M. Zinkevich and F. Aldinger, 'The Influence of the Solid Electrolyte on the Impedance of Hydrogen Oxidation at Patterned Ni Electrodes', in *Solid State Ionics* 181-25-26, p. 1170-1177, 2010.

[38] R. D. Armstrong and B. R. Horrocks, 'The Double Layer Structure at the Metal-Solid Electrolyte Interface', in *Solid State Ionics* 94, p. 181-187, 1997.

[39] S. Gewies and W. G. Bessler, 'Physically Based Impedance Modeling of Ni/YSZ Cermet Anodes', in *Journal of the Electrochemical Society* 155-9, p. B937-B952, 2008.

[40] M. G. H. M. Hendriks, J. E. ten Elshof, H. J. M. Bouwmeester and H. Verweij, 'The Electrochemical Double-Layer Capacitance of Yttria-Stabilised Zirconia', in *Solid State Ionics* 146-3-4, p. 211-217, 2002.

[41] N. L. Robertson and J. N. Michaels, 'Double-Layer Capacitance of Porous Platinum-Electrodes in Zirconia Electrochemical-Cells', in *Journal of the Electrochemical Society* 138-5, p. 1494-1499, 1991.

[42] S. Primdahl and M. Mogensen, 'Limitations in the Hydrogen Oxidation Rate on Ni/YSZ Anodes', in *Proceedings of the Sixth International Symposium on Solid Oxide Fuel Cells (SOFC-VI)*, p. 530-540, 1999.

[43] K. Sasaki, K. Susuki, A. Iyoshi, M. Uchimura, N. Imamura, H. Kusaba, Y. Teraoka, H. Fuchino, K. Tsujimoto, Y. Uchida and N. Jingo, 'H2S Poisoning of Solid Oxide Fuel Cells', in *Journal of the Electrochemical Society* 153-11, p. A2023-A2029, 2006.

[44] K. Eguchi, T. Setoguchi, K. Okamoto and H. Arai, 'An Investigation of Anode Material and Anodic Reaction for Solid Oxide Fuel Cell', in *ECS Transactions* 93-4, p. 494-503, 1993.

[45] T. Setoguchi, K. Okamoto, K. Eguchi and H. Arai, 'Effects of Anode Material and Fuel on Anodic Reaction of Solid Oxide Fuel-Cells', in *Journal of the Electrochemical Society* 139-10, p. 2875-2880, 1992.

[46] R. J. Aaberg, R. Tunold, S. Tjelle and R. Odegard, 'Oxidation of CO and H_2 on Ni / YSZ Cermet Electrodes', in *Proceedings of the 17th Risoe International Symposium on Materials Science: High Temperature Electrochemistry: Ceramics and Metals*, p. 511-516, 1996.

[47] F. Z. Boulenouar, K. Yashiro, M. Oishi, A. Kaimai, Y. Nigara, T. Kawada and J. Mizusaki, 'Electrochemical Oxidation of CO in a CO-CO2 System at the Interface of Ni Grid Electrode/YSZ Electrolyte', in *Electrochemical Society Series* 2001-16, p. 759-768, 2001.

[48] P. Holtappels, L. G. J. de Haart, U. Stimming, I. C. Vinke and M. Mogensen, 'Reaction of CO/CO2 Gas Mixtures on Ni-YSZ Cermet Electrodes', in *Journal of Applied Electrochemistry* 29-5, p. 561-568, 1999.

[49] G. O. Lauvstad, R. Tunold and S. Sunde, 'Electrochemical Oxidation of CO on Pt and Ni Point Electrodes in Contact With an Yttria-Stabilized Zirconia Electrolyte - II. Steady-State and Impedance Measurements', in *Journal of the Electrochemical Society* 149-12, p. E506-E514, 2002.

[50] A. Weber, B. Sauer, A. C. Müller, D. Herbstritt and E. Ivers-Tiffée, 'Oxidation of H-2, CO and Methane in SOFCs With Ni/YSZ-Cermet Anodes', in *Solid State Ionics* 152, p. 543-550, 2002.

[51] A. Utz, H. Störmer, D. Gerthsen, A. Weber and E. Ivers-Tiffée, 'Microstructure Stability Studies of Ni Patterned Anodes for SOFC', in *Solid State Ionics* 10.1016/j.ssi.2010.05.004, 2010.

[52] A. Leonide, S. Ngo Dinh, A. Weber and E. Ivers-Tiffée, 'Performance Limiting Factors in Anode Supported SOFC', in *Proceedings of the 8th European Solid Oxide Fuel Cell Forum*, p. A0501, 2008.

[53] A. Leonide, S. Hansmann, A. Weber and E. Ivers-Tiffée, 'Performance Simulation of C/V-Characteristics for ASC by Means of Detailed Impedance Analysis', in *Journal of Power Sources* 10.1016/j.powsour.2010.10.052, 2010.

[54] D. G. Goodwin, H. Y. Zhu, A. M. Colclasure and R. J. Kee, 'Modeling Electrochemical Oxidation of Hydrogen on Ni-YSZ Pattern Anodes', in *Journal of the Electrochemical Society* 156-9, p. B1004-B1021, 2009.

[55] A. Bieberle and L. J. Gauckler, 'State-Space Modeling of the Anodic SOFC System Ni, H2-H2O, YSZ', in *Solid State Ionics* 146-1-2, p. 23-41, 2002.

[56] W. G. Bessler, J. Warnatz and D. G. Goodwin, 'The Influence of Equilibrium Potential on the Hydrogen Oxidation Kinetics of SOFC Anodes', in *Solid State Ionics* 177-39-40, p. 3371-3383, 2007.

[57] M. Vogler and W. G. Bessler, 'The Role of Interstitial Hydrogen Species in Ni/YSZ Patterned Anodes: A 2D Modeling Study', in *ECS Transactions* 25, p. 1957-1966, 2009.

[58] A. Utz, H. Störmer, A. Leonide, A. Weber and E. Ivers-Tiffée, 'Degradation and Relaxation Effects of Ni Patterned Anodes in H_2/H_2O Atmosphere', in *Journal of the Electrochemical Society* 157-6, p. B920-B930, 2010.

[59] P. Holtappels, I. C. Vinke, L. G. J. de Haart and U. Stimming, 'Reaction of Hydrogen/Water Mixtures on Nickel-Zirconia Cermet Electrodes - II. AC Polarization Characteristics', in *Journal of the Electrochemical Society* 146-8, p. 2976-2982, 1999.

[60] H. Y. Zhu, R. J. Kee, V. M. Janardhanan, O. Deutschmann and D. G. Goodwin, 'Modeling Elementary Heterogeneous Chemistry and Electrochemistry in Solid-Oxide Fuel Cells', in *Journal of the Electrochemical Society* 152-12, p. A2427-A2440, 2005.

[61] V. Yurkiv, H.-R. Volpp and W. G. Bessler, 'Heterogeneous Chemistry and Electrochemistry of Carbon Monoxide at a Ni/YSZ Anode', in *Proceedings of the 9th EUROPEAN SOFC FORUM*, p. 9-1-9-14, 2010.

[62] T. H. Etsell and S. N. Flengas, 'Overpotential Behavior of Stabilized Zirconia Solid Electrolyte Fuel Cells', in *Journal of the Electrochemical Society* 118-12, p. 1890-&, 1971.

[63] M. Mogensen und K. V. Hansen, 'Impact of Impurities and Interface Reaction on Electrochemical Activity', in *Handbook of Fuel Cells - Fundamentals, Technology and Applications*, John Wiley & Sons Ltd, Chichester, p. 543-554, 2009.

[64] S. P. S. Badwal und J. Drennan, 'Interfaces in Zirconia Based Electrochemical Systems and Their Influence on Electrical Properties', in *Science of Ceramic Interfaces II*, Elsevier Science B.V., Amsterdam, p. 71-111, 1994.

[65] K. V. Hansen, K. Norrman and M. Mogensen, 'H2-H2O-Ni-YSZ Electrode Performance - Effect of Segregation to the Interface', in *Journal of the Electrochemical Society* 151-9, p. A1436-A1444, 2004.

[66] M. Mogensen, K. V. Jensen, M. J. Jorgensen and S. Primdahl, 'Progress in Understanding SOFC Electrodes', in *Solid State Ionics* 150-1-2, p. 123-129, 2002.

[67] J. Hogh, 'Influence of Impurities on the H2/H2O/Ni/YSZ Electrode', PhD Thesis, Risoe National Laboratory, Roskilde, Denmark, 2005.

[68] K. V. Jensen, 'The Ni-YSZ Interface: Structure, Composition and Electrochemical Properties at 1000 °C', PhD Thesis, Risoe National Laboratory, Roskilde, Denmark, 2002.

[69] M. Mogensen, S. Sunde and S. Primdahl, 'SOFC Anode Kinetics', in *High Temperature Electrochemistry: Ceramics and Metals*, p. 77-100, 1996.

[70] M. S. Schmidt, K. V. Hansen, K. Norrman and M. Mogensen, 'Effects of Trace Elements at the Ni/ScYSZ Interface in a Model Solid Oxide Fuel Cell Anode', in *Solid State Ionics* 179-27-32, p. 1436-1441, 2008.

[71] M. S. Schmidt, 'Effect of Dopants and Trace Elements at the Ni/ScYSZ Interface', PhD Thesis, University of Southern Denmark, Denmark, 2008.

[72] M. S. Schmidt, K. V. Hansen, K. Norrman and M. Mogensen, 'Three-Phase-Boundary Dynamics at the Ni/ScYSZ Interface', in *Solid State Ionics* 180-4-5, p. 431-438, 2009.

[73] K. Norrman, K. V. Hansen and M. Mogensen, 'Time-of-Flight Secondary Ion Mass Spectrometry As a Tool for Studying Segregation Phenomena at Nickel-YSZ Interfaces', in *Journal of the European Ceramic Society* 26-6, p. 967-980, 2006.

[74] A. E. Hughes, 'Segregation in Single-Crystal Fully Stabilized Yttria-Zirconia', in *Journal of the American Ceramic Society* 78-2, p. 369-378, 1995.

[75] M. de Ridder, A. G. J. Vervoort, R. G. van Welzenis and H. H. Brongersma, 'The Limiting Factor for Oxygen Exchange at the Surface of Fuel Cell Electrolytes', in *Solid State Ionics* 156-3-4, p. 255-262, 2003.

[76] E. Seddon, E. J. Tippett and W. E. S. Turner, 'The Electrical Conductivity of Sodium Meta-Silicate-Silica Glasses', in *Journal of the Society of Glass Technology* 16, p. 450-477, 1932.

[77] A. Utz, H. Störmer, T. Knapp, K. Ilin, P.-J. Jakobs, M. Siegel and E. Ivers-Tiffee, 'A Comparative Study of Fabrication Procedures for Ni-Patterned Anodes', in *Proceedings of the 9th EUROPEAN SOFC FORUM*, p. 5-91-5-103, 2010.

[78] H. J. Levinson und W. H. Arnold, 'Optical Lithography', in *Handbook of Microlithography, Micromachining and Microfabrication*, SPIE - The International Society of Optical Engineering, Bellingham, WA, p. 11, 1997.

[79] 'SPIP Scanning Probe Image Processor', in -4.0 1998.

[80] S. B. Adler, 'Reference Electrode Placement in Thin Solid Electrolytes', in *Journal of the Electrochemical Society* 149-5, p. E166-E172, 2002.

[81] E. Ivers-Tiffée, H. Timmermann, A. Leonide, N. H. Menzler und J. Malzbender, 'Methane Reforming Kinetics, Carbon Deposition, and Redox Durability of Ni/8 Yttria-Stabilized Zirconia (YSZ) Anodes', in *Handbook of Fuel Cells - Fundamentals, Technology and Applications*, John Wiley & Sons Ltd, Chichester, p. 933-956, 2009.

[82] 'MALT - MAterials-Oriented Little Thermodynamic Database', in -1.0 2004.

[83] H. Schichlein, A. C. Müller, M. Voigts, A. Krügel and E. Ivers-Tiffée, 'Deconvolution of Electrochemical Impedance Spectra for the Identification of Electrode Reaction Mechanisms in Solid Oxide Fuel Cells', in *Journal of Applied Electrochemistry* 32-8, p. 875-882, 2002.

[84] S. Primdahl and M. Mogensen, 'Gas Diffusion Impedance in Characterization of Solid Oxide Fuel Cell Anodes', in *Journal of the Electrochemical Society* 146-8, p. 2827-2833, 1999.

[85] J. R. Macdonald, *Impedance Spectroscopy*, John Wiley & Sons, New York, 1987.

[86] E. J. Opila, 'Oxidation and Volatilization of Silica Formers in Water Vapor', in *Journal of the American Ceramic Society* 86-8, p. 1238-1248, 2003.

[87] A. Utz, K. V. Hansen, K. Norrman, E. Ivers-Tiffée and M. Mogensen, 'Impurity Features in Ni-YSZ-H2-H2O Electrodes', in *Solid State Ionics* 10.1016/j.ssi.2010.12.014, 2010.

[88] R. B. Mclellan and W. A. Oates, 'Solubility of Hydrogen in Rhodium, Ruthenium, Iridium and Nickel', in *Acta Metallurgica* 21-3, p. 181-185, 1973.

[89] C. Wagner, 'Solubility of Vapor in Zro2-Y203 Mixed Crystals', in *Berichte der Bunsen-Gesellschaft fur Physikalische Chemie* 72-7, p. 778-781, 1968.

[90] A. Leonide, Y. Apel and E. Ivers-Tiffee, 'SOFC Modeling and Parameter Identification by Means of Impedance Spectroscopy', in *ECS Transactions* 19-20, p. 81-109, 2009.

[91] A. J. Bard and L. R. Faulkner, *Electrochemical Methods: Fundamentals and Applications*, John Wiley & Sons, New York, 2001.

[92] H. Wendt and G. Kreysa, *Electrochemical Engineering*, Springer, Berlin, 1999.

[93] S. Skaarup, B. Zachau-Christiansen and T. Jacobsen, 'Surface Species and Surface Mobility on Ni/YSZ Anodes', in *High Temperature Electrochemistry: Ceramics and Metals*, p. 423-430, 1996.

[94] W. G. Bessler, 'A New Computational Approach for SOFC Impedance From Detailed Electrochemical Reaction-Diffusion Models', in *Solid State Ionics* 176-11-12, p. 997-1011, 2005.

[95] T. Yamamura, H. Yoshitake, H. Tagawa, N. Mori, K. Hirano, J. Mizusaki, S. Ehara, T. Takagi, M. Hishinuma, H. Sasaki, Y. Nakamura and K. Hashimoto, 'Experimental Evidence for Three Phase Boundary As Active Site on Nickel / YSZ System', in *Proceedings of the 2nd European Solid Oxide Fuel Cell Forum* 2, p. 617-625, 1996.

[96] C. H. Hamann, A. Hamnett and W. Vielstich, *Electrochemistry*, 2. edition, Wiley-VCH, Weinheim, 2007.

[97] C. Peters, *Grain-Size Effects in Nanoscaled Electrolyte and Cathode Thin Films for Solid Oxide Fuel Cells (SOFC)*, Universitätsverlag Karlsruhe, Karlsruhe, 2009.

[98] A. Leonide, S. Hansmann and E. Ivers-Tiffee, 'Performance Simulation of C/V Characteristics for ASC by Means of Detailed Impedance Analysis', in *Proceedings of the 9th EUROPEAN SOFC FORUM*, p. 6-1-6-6, 2010.

[99] A. Leonide, 'SOFC Modelling and Parameter Identification by means of Impedance Spectroscopy', Dissertation, Karlsruher Institut für Technologie (KIT), Karlsruhe, Germany, 2010.

[100] W. G. Bessler, S. Gewies and M. Vogler, 'A New Framework for Physically Based Modeling of Solid Oxide Fuel Cells', in *Electrochimica Acta* 53-4, p. 1782-1800, 2007.

[101] J. Bisquert, G. G. Belmonte, F. F. Santiago, N. S. Ferriols, M. Yamashita and E. C. Pereira, 'Application of a Distributed Impedance Model in the Analysis of Conducting Polymer Films', in *Electrochemistry Communications* 2-8, p. 601-605, 2000.

[102] G. Paasch, K. Micka and P. Gersdorf, 'Theory of the Electrochemical Impedance of Macrohomogeneous Porous-Electrodes', in *Electrochimica Acta* 38-18, p. 2653-2662, 1993.

[103] P. R. Shearing, J. Golbert, R. J. Chater and N. P. Brandon, '3D Reconstruction of SOFC Anodes Using a Focused Ion Beam Lift-Out Technique', in *Chemical Engineering Science* 64-17, p. 3928-3933, 2009.

[104] J. R. Wilson, W. Kobsiriphat, R. Mendoza, H. Y. Chen, J. M. Hiller, D. J. Miller, K. Thornton, P. W. Voorhees, S. B. Adler and S. A. Barnett, 'Three-Dimensional Reconstruction of a Solid-Oxide Fuel-Cell Anode', in *nature materials* 5-7, p. 541-544, 2006.

[105] P. S. Jorgensen, K. V. Hansen, R. Larsen and J. R. Bowen, 'A Framework for Automatic Segmentation in Three Dimensions of Microstructural Tomography Data', in *Ultramicroscopy* 110-3, p. 216-228, 2010.

[106] B. Rüger, J. Joos, T. Carraro, A. Weber and E. Ivers-Tiffée, '3D Electrode Microstructure Reconstruction and Modelling', in *ECS Transactions* 25-2, p. 1211-1220, 2009.

[107] B. Rüger, A. Weber and E. Ivers-Tiffée, '3D-Modelling and Performance Evaluation of Mixed Conducting (MIEC) Cathodes', in *ECS Transactions* 7, p. 2065-2074, 2007.

[108] B. Rüger, *Mikrostrukturmodellierung Von Elektroden Für Die Festelektrolytbrennstoffzelle*, Universitätsverlag Karlsruhe, Karlsruhe, 2009.

[109] B. Butz, R. Schneider, D. Gerthsen, M. Schowalter and A. Rosenauer, 'Decomposition of 8.5 Mol.% Y2O3-Doped Zirconia and Its Contribution to the Degradation of Ionic Conductivity', in *Acta Materialia* 57-18, p. 5480-5490, 2009.

[110] B. Butz, P. Kruse, H. Störmer, D. Gerthsen, A. C. Müller, A. Weber and E. Ivers-Tiffée, 'Correlation Between Microstructure and Degradation in Conductivity for Cubic Y_2O_3-Doped ZrO_2', in *Solid State Ionics* 177-37-38, p. 3275-3284, 2006.

[111] V. Sonn and E. Ivers-Tiffée, 'Degradation in Ionic Conductivity of Ni/YSZ Anode Cermets', in *Proceedings of the 8th European Solid Oxide Fuel Cell Forum*, p. B1005, 2008.

[112] M. Vogler, H. Störmer, D. Gerthsen, A. Utz, A. Weber and E. Ivers-Tiffee, 'Electrochemistry and Mechanism of Hydrogen Oxidation at Ni/YSZ Patterned Anodes', in *Proceedings of the 9th EUROPEAN SOFC FORUM*, p. 9-94-9-103, 2010.

[113] A. Utz, A. Weber, E. Ivers-Tiffée and H. Störmer, 'Studies on Pattern Anode Stability for SOFC', in *Proceedings of the 8th European Solid Oxide Fuel Cell Forum*, p. A0509-A0519, 2008.

[114] A. Utz, A. Leonide, A. Weber and E. Ivers-Tiffee, 'Studying the CO/CO2 Characteristics of SOFC Anodes by Means of Patterned Ni Anodes', in *Journal of Power Sources* 10.1016/j.powsour.2010.10.056, 2010.

[115] J. Winkler, P. V. Hendriksen, N. Bananas and M. Mogensen, 'Geometric Requirements of Solid Electrolyte Cells With a Reference Electrode', in *Journal of the Electrochemical Society* 145-4, p. 1184-1192, 1998.

[116] R. Männer, E. Ivers-Tiffée and W. Wersing, 'Characterization of YSZ Electrolyte Materials With Various Yttria Contents', in *Proceedings of the Second International Symposium on Solid Oxide Fuel Cells (SOFC-II)*, p. 715-725, 1991.

[117] K. Thydén, 'Microstructural Degradation of Ni-YSZ Anodes for Solid Oxide Fuel Cells', PhD Thesis, Risoe National Laboratory for Sustrainable Energy, Technical University of Denmark, Roskilde, Denmark, 2008.